Mass Growth of Large $PbWO_4$ Single Crystals for Particle Detection in High-Energy Physics Experiments at CERN

Mass Growth of Large $PbWO_4$ Single Crystals for Particle Detection in High-Energy Physics Experiments at CERN

A. N. Annenkov
Yu. S. Kuz'minov

Viva Books

New Delhi | Mumbai | Chennai | Kolkata | Bengaluru | Hyderabad | Kochi | Guwahati

First Indian Edition 2012

VIVA BOOKS PRIVATE LIMITED

- 4737/23 Ansari Road, Daryaganj, New Delhi 110 002
  E-mail: vivadelhi@vivagroupindia.net, Tel. 42242200
- 76, Service Industries, Shirvane, Sector 1, Nerul, Navi Mumbai 400 706
  E-mail: vivamumbai@vivagroupindia.net, Tel. 27721273, 27721274
- Jamals Fazal Chambers, 26 Greams Road, Chennai 600 006
  E-mail: vivachennai@vivagroupindia.net, Tel. 28290304, 28294241
- B-103, Jindal Towers, 21/1A/3 Darga Road, Kolkata 700 017
  E-mail: vivakolkata@vivagroupindia.net, Tel. 22836381, 22816713
- 7, Sovereign Park Apartments, 56-58, K. R. Road, Basavanagudi, Bangalore 560 004
  E-mail: vivabangalore@vivagroupindia.net, Tel. 26607409, 26607410
- 101-102 Mughal Marc Apartments, 3-4-637 to 641, Narayanguda, Hyderabad 500 029
  E-mail: vivahyderabad@vivagroupindia.net, Tel. 27564481, 27564482
- Beevi Towers, First Floor, SRM Road, Kaloor, Kochi 682 018, Kerala
  E-mail: vivakochi@vivagroupindia.net, Tel: 0484-2403055, 2403056
- 232, GNB Road, Beside UCO Bank, Silpukhuri, Guwahati 781 003
  E-mail: vivaguwahati@vivagroupindia.net, Tel: 0361-2666386

www.vivagroupindia.com

Published by arrangement with
Cambridge International Science Publishing
7, Meadow Walk, Great Abington
Cambridge CB1 6AZ, UK

ISBN : 978-81-309-1992-8

Published by Vinod Vasishtha for Viva Books Private Limited, 4737/23, Ansari Road, Daryaganj, New Delhi-110 002. Printed and bound by Syndicate Binders, Noida.

Contents

Preface

The book discusses the problems of growing not one or several but many thousands of highly perfect and relatively large crystals (for example, 230 mm long, with a diameter of 40 mm). These crystals are the subject of this book. They were required for investigating global problems of the structure and formation of our universe. Appropriate equipment was needed for solving these problems. The problem of growth of the crystal appears to be secondary. Although, like the heart in the human organism determines health, the quality of the crystal solves many problems in detection systems. The task is to grow a very large number of crystals without losing their valuable properties. For this purpose, it was necessary to reorganise the entire growth process, postgrowth cooling, annealing and solve a number of small and large problems. However, it should be said that there are no small problems in crystal growth. A small error may result in considerable damage and make a crystal unsuitable and it is therefore necessary to repeat the entire growth process. Therefore, developers and technologists try to automate fully the entire growth process in order to exclude the so-called human factor. However, there are many other tasks. The crystal must be checked in every stage. If the data do not fit the required range, the crystal is remelted.

The methods used for investigating crystals should be relatively simple but highly accurate. Of considerable importance is the statistical processing of the investigation results which makes it possible to predict the proportion of crystals with unacceptable defects. The method and equipment used in these investigations can also be applied in other areas of production of scarce and important components in electronics, medicine and other areas. After all, automation can help greatly in solving technical problems, but the main responsibility rests with people who should rally to solve this problem. CERN is a suitable example here. At present, approximately 7000 scientists from 50 countries work there. In most cases, the work is of the collected nature and such common work strengthens of the team. Such a team was also formed in the development of the electromagnetic calorimeter for the CMS experiments on the LHC.

We would like to wish scientists and technicians every success in completing this work and carrying out the planned experiments. This will represent a significant contribution to investigating the mysteries of our universe.

I.A. Shcherbakov

Director of the General Physics Institute, Russian Academy of Sciences
Corresponding Member of the Russian Academy of Sciences

Introduction

Experiments in high energy physics, planned for the next decade in the leading world centres investigating the macroworld, such as the CERN where the most advanced LHC accelerator has been constructed, require the development of appropriate systems for detecting and identifying particles. One of these systems uses electromagnetic calorimeters based on scintillation crystals. Inorganic scintillation crystals represent the basis of the most advanced measurement systems and provide extensive possibilities for obtaining results which should prove to be revolutionary in fundamental physics. High energy physics uses experimental systems with very large number of scintillators.

For example, the calorimeter of the CMS experiment (Compact Muon Solenoid) must ensure the unique energy resolution in recording γ-quanta. It should be better than $2.5\%/\sqrt{E}$, where E is energy of γ-quanta in GeV. The expected service life of the calorimeter is no less than 10 years. This means that stringent requirements are imposed on the technical parameters of the scintillators and variations of their properties. The $PbWO_4$ scintillation crystals (lead tungstate, PWO), used in the detector cells, should be radiation hard, emit at least 90% of light in the first 100 ns, their light yield should be greater than 8 photoelectrons/MeV, and the nonuniformity of the light field along the length of the crystals should not exceed several percent.

The book deals with the scientific and practical aspects of solving the target problem – the development of the most advanced technology and introduction into mass production of a new generation of lead tungstate scintillation crystals for application in experiments in accelerators with high luminosity. The investigations of the scintillation mechanism in the tungstate crystals and their relationship with the technological factors represent the scientific fundamentals of the development of the technology of directional growth of crystals from the melt by the Czochralski method for mass production.

1. Application of crystalline scintillators in high-energy physics

1.1. Scintillators

Scintillators are substances which emit photons in the visible or ultraviolet part of the spectrum under the effect of charged particles or electromagnetic radiation. The emission spectrum of electromagnetic radiation is shifted in relation to the absorption spectrum.

The development of scintillation technology for the recording of ionising radiation has made it possible to transfer in physical investigations from empirical, qualitative consideration to systematic quantitative analysis of the processes taking place in the microworld. The beginning of experimental nuclear physics is represented the experiments carried out by Rutherford in the investigation of scintillations in ZnS crystals, bombarded with α-particles. The first scintillation material, used for recording x-ray radiation, was calcium tungstate $CaWO_4$.

Only several scintillator types were used up to the middle of the 40s of the previous century. However, the development of atomic technologies at the end of the 40s of the 20th century resulted in a rapid increase of the interest in the methods of recording ionising radiation, including by scintillation counters. It was found that scintillations can be detected in various organic and inorganic crystalline compounds [1–4], in liquids [5–8], in gases [9, 10] and in polymer compounds [11]. Consequently at the beginning of the 50s, the main classes of inorganic crystalline compounds which were promising for investigations of scintillations were determined and the scintillation crystalline material NaI (Tl) was discovered which has been used most extensively up to now [12].

Excluding inert gas and some special gaseous compounds, the inorganic scintillators are represented in most cases by single crystals or polycrystalline powders. The single crystals are usually transparent for intrinsic luminescence, whereas powders are less transparent

The largest group of the single crystal scintillators consists of alkali halogenides, mainly with additions of metals, such as: thallium, europium and sodium, and also pure and doped oxide compounds, For example, $NaI(Tl^{+})$, $CsI(Tl^{+})$, $CsI(Na^{+})$, $CaWO_4$, $CdWO_4$, BaF_2, $Bi_4Ge_3O_{12}$, $Gd_2SiO_5(Ce^{3+})$, $Lu_2SiO_5(Ce^{3+})$ and $YAlO_3(Ce^{3+})$.

It is interesting that the search for and development of new laser media based on fluoride and oxygen in organic compounds in the 60s of the 20th century stimulated extensive investigations into different inorganic crystalline materials during which promising scintillators were also discovered.

In the 80s of the 20th century, after discovering bismuth trigermanate (BGO) and the start of extensive application in medical diagnostics and experimental physics, a completely new direction was formed for the search for new scintillation materials amongst oxygen-containing compounds. As a result, scintillation crystalline materials based on silicates of gadolinium and lutecium, activated with trivalent cerium ions were discovered at the end of the 80s and used extensively in devices for medical diagnostics. At the beginning of the 90s, as a result of efficient research, lead tungstate and lutecium–aluminium perovskite, activated with cerium, were discovered. They were used in high-energy physics, medical diagnostics and industry. The dynamics of the development of inorganic scintillation materials is shown in Fig. 1.1 [13].

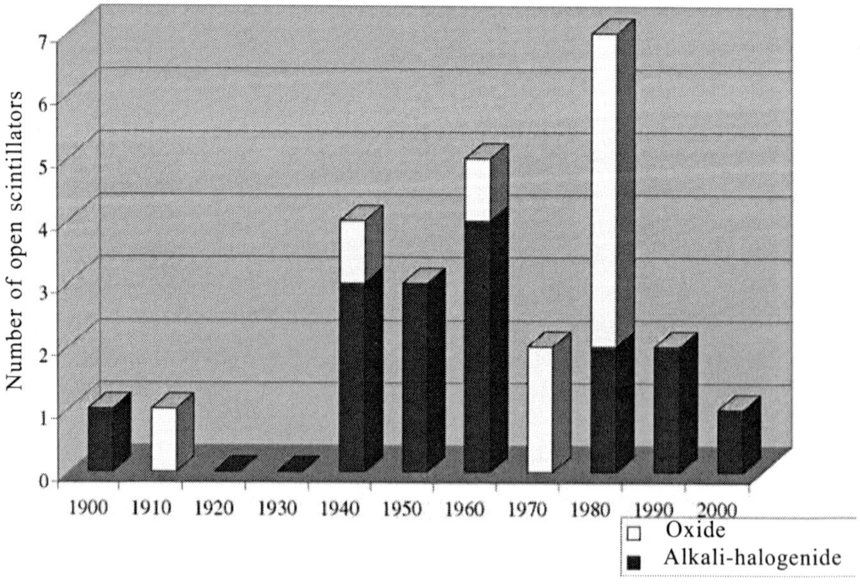

Fig. 1.1. Dynamics of development of inorganic scintillation materials.

In this book, attention is only given to aspects of inorganic crystalline scintillators. In inorganic crystals, luminescence forms in the emission centres of the compounds or is produced by activating additions added in specific amounts to the crystal lattice.

The band model is used to explain the mechanism of luminescence of inorganic crystals. In most cases, scintillation crystals are dielectrics, less frequently semiconductors, with the width of the forbidden band of several electron volts. In interaction with ionising radiation or a charged article, part of the energy is transferred to the crystal lattice of the scintillators and transfers the electrons from the valence band into the conduction band (excitation), leaving holes in the valence band. The electrons or holes are transferred in the crystal to activation centres in which recombination processes with photon emission take place.

However, this description of the process applies only to perfect crystals. Lattice imperfections, formed as a result of growth defects or the presence of an impurity in the crystal, generate additional levels in the forbidden band and excitons or electrons from the conduction band fall on these levels. These impurity levels form activation centres of three types [14]:

- luminescence centres in which recombination of a pair takes place through the excited state to the ground state with photon emission. The electron centre is transferred into the excited state as a result of capture of an exciton or random capture of an electron from the conduction band and a hole from the valence band;
- decay centres – these are similar to luminescence centres with the only difference being that in transition from the excited state to the ground state the energy is transferred into the photon spectrum instead of light emission;
- traps – these are metastable levels on which the electrons and holes or excitons can exist for a relatively long time until the thermal or other energy becomes sufficient to return them to their bands moving to the luminescence or decay centre. These processes are characterised by the indeterminacy of the emission time and are referred to as phosphorescence.

The wide range and successful application of the crystalline scintillation detectors have been based mainly on the following properties:

- considerably higher (in comparison with gas-discharge counters) efficiency of recording (high degree of absorption in the scintillators) of x-ray radiation, α- and β-particles, especially in recording of high-energy γ-radiation (from hundreds of keV to hundreds of GeV);

- the decay time of scintillation is in the range from hundreds of nanoseconds to microseconds, ensuring a high count rate and high time resolution power and enabling measurements of the elements to be taken in the time range up to tenths of a nanosecond;
- the almost completely linear scintillation efficiency over the entire width of the recorded energy spectrum so that particles can be separated;
- efficient matching of the scintillation spectrum of the crystal with the spectral sensitivity of the most extensively used photoelectric multipliers;
- the variable geometrical form and volume offer extensive possibilities for carrying out experiments in specific conditions.
- at present, the scintillators and devices based on them are used widely in many areas of science, technology and in household devices. The main areas of application will be mentioned briefly [15, 16]: nuclear medicine, flaw detection, oil prospecting, dosimetry and radiometry, radio astronomy, fundamental scientific investigations, safety and monitoring of radioactive substances.

In fundamental investigations, the scintillators are used for constructing calorimeters and detectors of various type designed, for example, for detecting and investigating elementary particles.

Various areas of application impose different and, in most cases, contradicting requirements on the scintillators used. Therefore, it is not possible to produce a single universal scintillator. Scintilators are selected for specific applications on the basis of the radiation to be studied, the spectral characteristic of the photoelectric multiplier and some special requirements, for example, the short afterglow period or high pulse amplitude. Nevertheless, the quality of the scintillators (user properties) can be estimated on the basis of several generalising criteria.

To obtain a high transformation coefficient, the degree of absorption of primary radiation should be high. Therefore, to record high energy penetrating radiation, it is necessary to select scintillators with a higher density and a high effective atomic number *Z*.

The losses of luminescence radiation as a result of reflection and absorption in the scintillator should be minimum.

The maximum possible part of the primary energy absorbed by the scintillator should be transformed to luminescence radiation. Kalman [2] refers to this coefficient as the physical yield. He also proposed a generalised estimate of the scintillator, taking into account the given requirements, and introduced the parameter 'technical yield'. This parameter is the number of photons emitted by the scintillator in the direction of the photocathode for the fixed intensity of the specific type of ionising radiation and at the optimum longitudinal dimension of the scintillating element. An increase of the physical yield is also accompanied by an increase of the technical yield.

As the degree of natural absorption of luminescence radiation in the scintillator decreases, the thickness of the scintillator should increase and the degree of absorption of primary radiation and, consequently, the technical yield should also increase.

The spectral composition of luminescence radiation should correspond to the spectral sensitivity of the photocathode of the multiplier.

For certain applications of scintillation sensors it is very important that the duration of the pulse should be as short as possible and the amplitude as high as possible. This is especially critical, for example, in medical tomography where the duration of irradiation of the biological object should be as short as possible. In order to obtain a high speed of light and high time resolution capacity, it is necessary to select a scintillator with the shortest luminescence and no afterglow.

When using the scintillation sensors for spectrometric measurements, there should be a proportionality between the energy absorbed in the scintillator and the number of the emitted photons.

In addition, the physical yield for highly ionised particles (for example, α-particles) is lower than for the particles with a lower ionising capacity (for example, β-radiation). This capacity is characterised by the ratio of the signals α/β and is also important when selecting the scintillator.

The physical properties of several widely used in organic scintillators are presented in Table 1.1 [14–16, 19].

1.2. Electromagnetic calorimetry in high-energy physics

High-energy physics studies processes and laws of nature relating to the deepest, primary levels of matter. The concept of 'the primary levels of matter' and also the concept of 'high energy' have been constantly changing with obtaining further information on the world and technologies which make it possible to transfer the problems of investigations of specific levels of matter from the area of speculative hypothesis to the region of experimentally verified facts.

Table 1.1. Characteristics of several inorganic scintillators

Property	NaI:Tl	YAG:Ce	YAP:Ce	BGO	CaF_2:Eu	CsI:Tl
Density, g/cm^3	3.67	4.55	5.35	7.13	3.18	4.51
Moos hardness	2	8,5	8,6	5	4	2
Crystal structure	Cubic	Cubic	Rhombic	Cubic	Cubic	Cubic
Melting point, °C	651	1970	1875	1050	1418	621
Hygroscopicity	Yes	No	No	No	No	Slight
Chemical formula	NaI	$Y_3Al_5O_{12}$	$YAlO_3$	$Bi_4Ge_3O_{12}$	CaF_2	CsI
Relative light yield, %	100	26	38	19	50	120
Radiation length, cm	2.56	3.28	2.2	1.12	3.72	2.43
Absolute light yield at 300 K, 10^3 photons/MeV	43	11	16.2	8.2	21.5	51.8
Luminescence spectrum maximum, nm	415	550	347	505	435	560
Scintillation luminescence time, ns	230	70	30	300	940	1000

The experimental investigations in high energy physics are not possible without the development of large scale equipment and systems using the latest scientific and technological developments and, correspondingly, requiring extremely high investment. High energy physics experiments are usually carried out on the basis of international collaboration, so that intellectual, technological and financial resources of many countries can be combined together. Therefore, experimental high energy physics is a suitable example of successful international cooperation for more than 50 years.

The foundation of the European Organisation for Nuclear Research (CERN) in 1954 for carrying out combined investigations in the area of high energy physics was one of the first joint European project. At present, 20 countries are official members of CERN. CERN combines more than 7000 scientists, working in more than 500 universities and research centres in more than 80 countries of the world. The members of CERN are: Austria, Belgium, Bulgaria, Czech Republic, Denmark, Finland, France, Germany, Greece, Hungary, Italy, Netherlands, Norway, Poland, Portugal, Slovak Republic, Spain, Sweden, Switzerland, Great

Britain. The status of observers in CERN has been awarded to: India, Turkey, Israel, Russia, Japan, USA, UNESCO and EC Commission (www.cern.ch).

Scientific equipment at CERN includes various particle accelerators: the proton synchrotron (PS), the superproton synchrotron (SPS), the recently constructed Large Hadron Collider (LHC) and detection systems for experiments. Each CERN system is a basis for carrying out several large expensive experiments. Figure 1.2 shows the aerial photograph of the region of the CERN [17].

The most important and noticeable event in recent years at the CERN was the construction of the Large Hadron Collider, which has lasted almost 20 years. In October 2008, this grandiose scientific system was activated resulting in considerable interest throughout the world. The LHC is the most powerful, unique and most expensive accelerator in the world. In the next 10 years, the LHC is to be used for four large experiments, two of which, CMS and ATLAS, are aimed at finding Higgs bosons and supersymmetry, and two others, LHCb and ALICE at investigating quark-gluon plasma. The successful results of this experiment will provide unique information on the structure of the universe. Figure 1.3 shows the locations of the experiments in the LHC accelerator [18].

Fig. 1.2. The aerial photograph of the region of the CERN. Circles indicate the underground location of the accelerators.

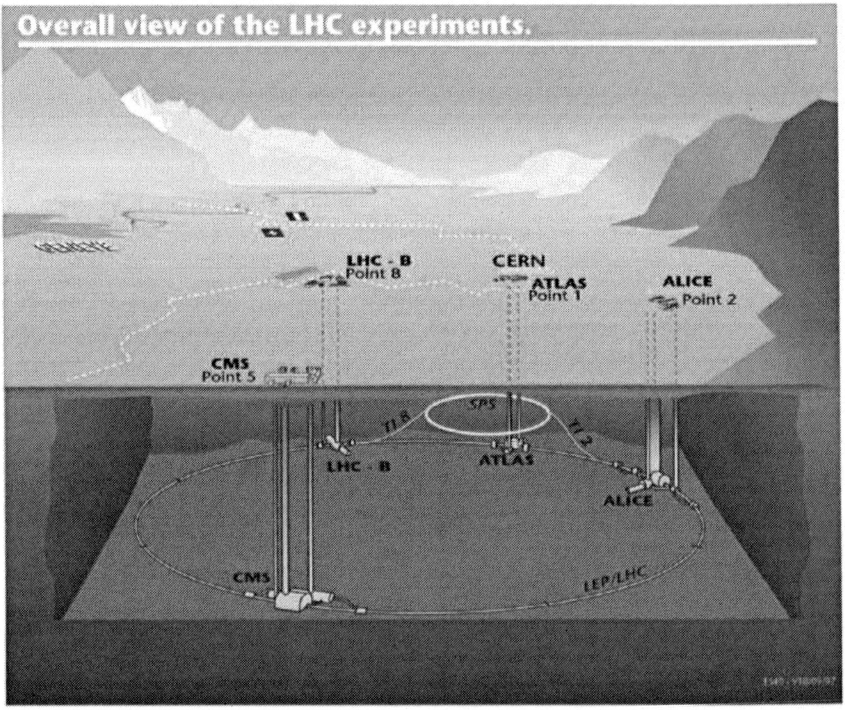

Fig. 1.3. Location of experiments in the LHC accelerator.

The central part of experimental systems in the LHC are electromagnetic calorimeters, i.e. complex systems which make it possible to reproduce with high accuracy the distribution of the field of γ-quanta formed in collision of particles in the Collider.

In the group of different types of electromagnetic calorimeters the special position is occupied by a homogeneous calorimeter based on scintillation crystals, capable of ensuring high energy and spatial resolution (Fig. 1.4) [20]. However, none of the scintillators developed and used up to the beginning of the 90s was suitable for the new generation of experiments in the LHC, where the frequency of collisions of the packets of the particles exceeds 40 MHz.

Figure 1.5 shows the general view of the CMS detector. The following main part of the detector can be seen: the magnetic and muonic systems; the track detector; the crystal electromagnetic high-resolution calorimeter with two end sections, the hadron calorimeter, and the very forward calorimeter. As a result of the large radius of the magnetic solenoid and compact design of the electromagnetic and hadron calorimeters of the CMS detector, all the calorimetric systems

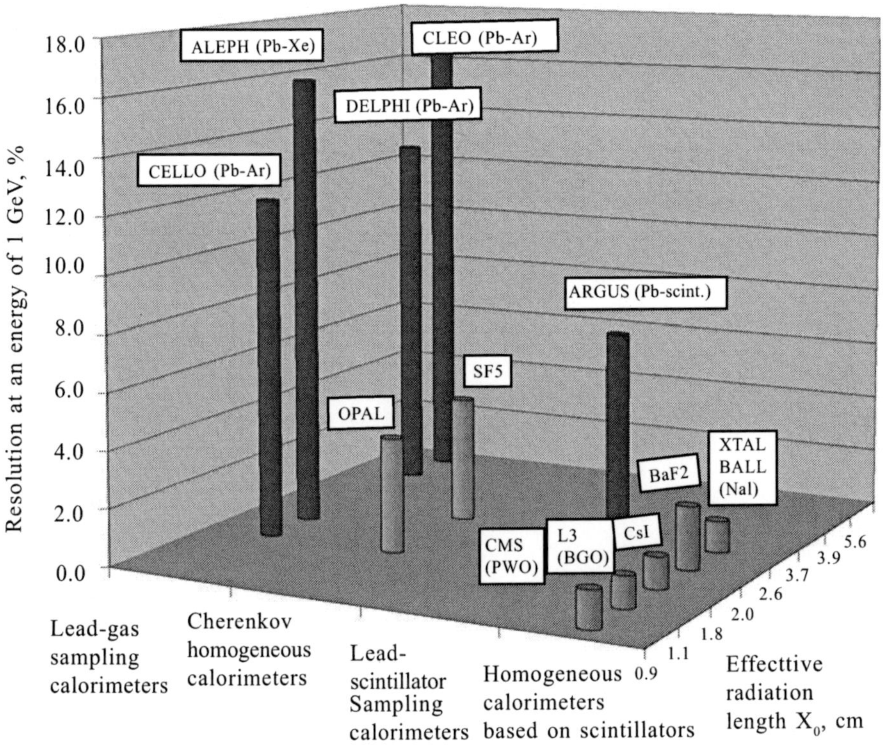

Fig. 1.4. Energy resolutions in complete absorption detectors (electromagnetic calorimeters) in different experiments.

are situated inside the solenoid so that high-resolution of the detector can be achieved.

The requirements on the electromagnetic calorimeter (ECAL) of the CMS project are determined by the need to analyse the processes of breakdown of Higgs bosons with intermediate mass. In this case, the critical parameter is the mass resolution of two-photon events [19]. Mass resolution depends on the energy resolution of two photons E_1, E_2 and angular divergence θ by means of the relationship:

$$\frac{\sigma_M}{M} = \frac{1}{2}\left[\frac{\sigma_{E_1}}{E_1} \oplus \frac{\sigma_{E_2}}{E_2} \oplus \frac{\sigma_\theta}{\tan(\theta/2)}\right] \tag{1.1}$$

where ⊕ is the sign of quadratic summation, E is measured in GeV, and θ in radians. Energy resolution is described by the equation:

$$\frac{\sigma_E}{E} = \left[\frac{a}{\sqrt{E}} \oplus \frac{\sigma_N}{E} \oplus c\right] \tag{1.2}$$

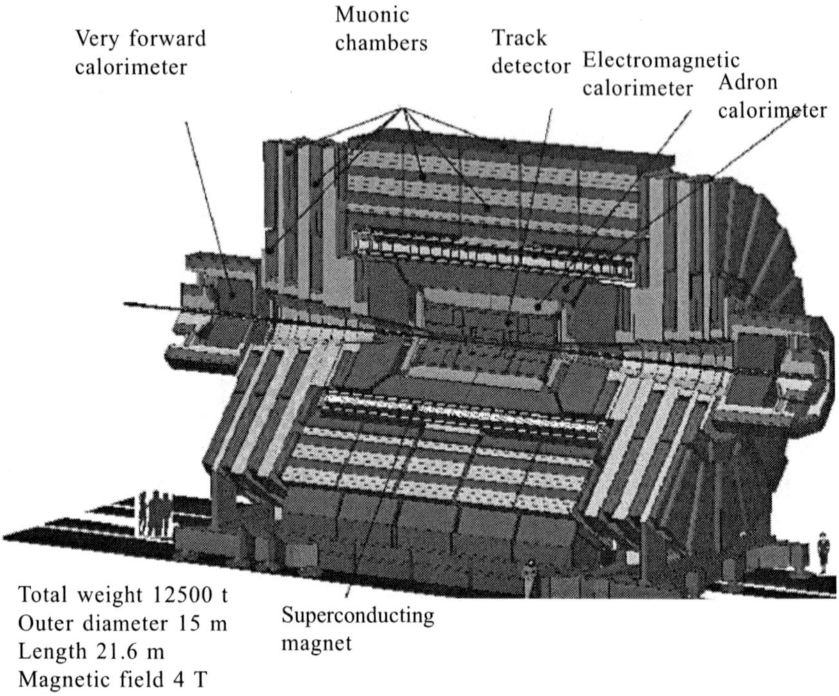

Fig. 1.5. General view of the detector for CMS experiments to find Higgs bosons [19].

where *a* is a stochastic term, σN is the energy equivalent of noise, and *c* is a constant. The noise component has two consttuets, electronic noise and signal overlapping. To obtain the highest energy resolution, the contribution of all components should be minimum and of the same order with the appropriate photon energies.

The following requirements were fulfilled when designing the electromagnetic calorimeter:

- optimum energy and angular resolution
- capacity for cutting of signals from π^0 mesons
- resistance to radiation loading from neutron radiation inside and around the calorimeter
- geometrical suitability, characterised by the limit of the speed range $|\eta|$ [19] and restricting the accuracy of energy measurements
- the optimum resolution with respect to mass at low and high luminosity levels.

Many years of investigations by scientists, engineers and technologists of several countries were completed in 1994 in CMS collaboration when it was decided to construct an electromagnetic calorimeter with high-resolution on the basis of lead tungstate crystals. The calorimeter consists of 76 000 crystalline elements with a total

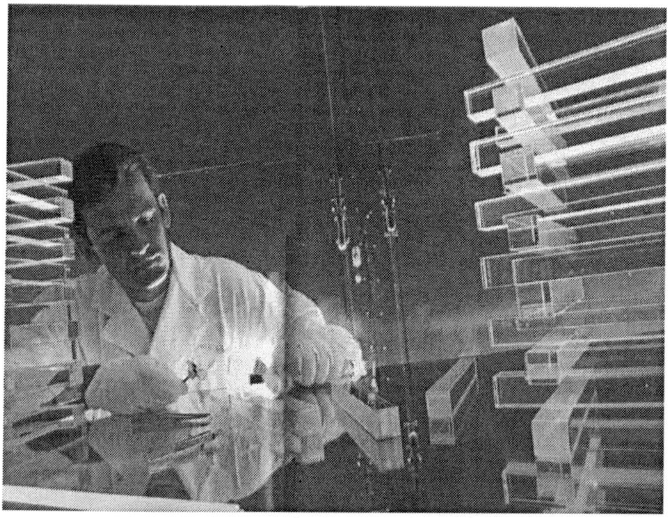

Fig. 1.6. PWO scintillators (CERN photo archive).

weight of approximately 90 tonnes. The process of certification of the scintillation elements is shown in Fig. 1.6.

1.3. Lead tungstate scintillation crystal

The lead tungstate crystals have been attracting attention of researchers for more than 60 years. However, advances in the development of technology for this material were made only in the middle of the 90s and this resulted in the production of large single crystals with acceptable scintillation parameters. The results of investigations carried out in early stages of the CMS project show that, in comparison with other scintillators, the PWO crystal has the highest density, the smallest radiation length X_0 and the Moliere radius R_M so that a compact electromagnetic calorimeter can be produced. Consequently, the lead tungstate has become a potentially promising material for high energy physics.

1.3.1. The structure of PWO crystals

It is well known that tungsten oxycations form with bivalent cations of the second and eighth group of the periodic system of chemical elements, and also with Cu, Mn and Pb, a number of compounds with a simple stoichiometric composition with the general formula AWO_4, where A = Mg, Ca, Zn, Sr, Cd, Co, Ni, Mn, Pb. As regards the structure, the series breaks down into tungstites (cation radius $r < 10^{-10}$ m) and scheelites ($r > 10^{-10}$ m). However, the change of the type of structure is not always determined only by the purely geometrical factor – cation radius. Lead tungstate (cation radius $1.2 \cdot 10^{-10}$ m) can crystallise both in the structural type of scheelite β–$PbWO_4$ (stolzite, as a result of the large radius), and tungstite α–$PbWO_4$ (raspite, as a result of the polarisation effect) [21, 22]. The main properties of pure structural types of PWO are presented in Table 1.2.

The lead tungstate with the tungstite structure has the two-layer densest packing of oxygen atoms in which half of the octahedral cavities is occupied by Pb atoms on the same level and by W atoms on the other level. In the cation base of the tungstite structure, the atoms of Pb and W occupy alternate planes. The octahedrons with the cations of the same type form zigzag chains and the isolated chains of Pb octahedrons form the perpendicular layer [100], alternating with the layer of the W-octahedrons which touch along the faces.

The scheelite structure is based on a three-dimensional frame, formed by infinite zigzag chains of the Pb polyhedrons. The Pb octahedrons are joined along side faces into spirals around quadruple helical axis parallel to [001]. Between each pair of the polyhedrons there are single orthotetrahedrons WO_4 not connected with each other. Thus, in the $PbWO_4$ morphotropic series the structural type of scheelite is evidently more stable (cation radius $>10^{-10}$ m). However, the difference in the

Table 1.2. Morphotropic structural types of lead tungstate

Property	α-$PbWO_4$	β-$PbWO_4$
Structure type	Wolframite	Scheelite
Crystal system	Monoclinic	Tetragonal
Spatial group	P2/C	I41/a
Lattice parameters, Å:		
a	5.58	5.456
b	5.00	5.456
c	13.64	12.020
β	107°33'	
Moos hardness	3.0	3.5 – 4.0
Density, g/cm^3	8.4 – 8.5	7.9 – 8.3
Clevage	Perfect on {100}	Imperfect on {001}
Optical sign	Uniaxial, negative	Uniaxial, negative
Refractive index:		
n_o		2.27
n_e		2.15

main structural types (tungstite and scheelite) is not absolute because the transition from one type to another is clearly visible. The change of the structural type of tungstite to the structural type of scheelite is accompanied by a small increase of the volume per oxygen atom. In other words, the structure of tungstite can be regarded as distortion of the scheelite structure.

It has been found [23] that the lead tungstate crystals are usually a solid solution of two phases – pure scheelite and a scheelite-like phase, with the main structural unit of such a solution being the tungsten oxycomplex WO_4^{2-} and bivalent lead ions or their vacancies in the octuple coordination. The fraction of every phase in the crystal depends on the growth conditions. In most cases, the ratio of the phases in the solid solution is not stabilised at a single value and changes from crystallisation to crystallisation. Point defects in the structure, found in lead tungstate crystals represent the sum of defects of two structural types of crystal. The scheelite crystal is characterised by the random distribution of vacancies caused by the low and stochastically distributed (over the crystal) shortage of PbO in crystal growth; in addition, the scheelite-like crystal also contains regular cation vacancies in one of the lead positions. The qualitative dependence of the concentration of vacancies on the crystal structure is shown schematically in Fig. 1.7 [24, 25].

This dependence reflects the fact that the growth of large PWO crystals by the Czochralski method from the initial stoichiometric melt is accompanied by the formation of a lead deficit in the melt causing the formation of randomly distributed cation vacancies V_c in the positions of the lead ions. These vacancies are compensated by the oxygen vacancies V_O which are also randomly distributed in the matrix.

1.3.2. Industrial methods of growing PWO crystals, analysis of the equilibrium diagram

The lead tungstate crystals are produced by crystallisation from a melt and the current industrial methods are the Czochralski and Bridgman methods. The growth methods have been described in detail in [26]. The present book deals with the peculiarities and relationships of the technology of growing PWO crystals by the Czochralski method.

The equilibrium diagram of the PbO–WO_3 system has been studied quite extensively by the methods of x-ray diffraction and DTA analysis. However, the diagram has been studied in greater detail on the side of low WO_3 concentrations. There has been only a small number of investigations in the area with a higher content of tungsten anhydride,

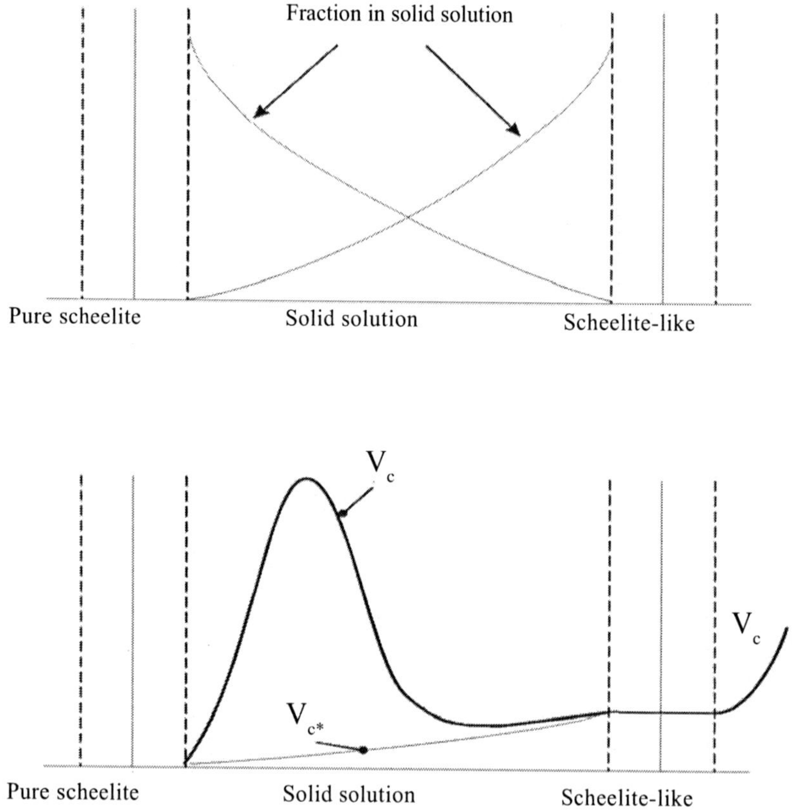

Fig. 1.7. Qualitative dependence of the number of cation vacancies on the crystal structure.

because of the high melting point of WO_3 and the energy interaction of WO_3 with the crucible materials in the melt at temperatures above 1100°C. It has been only established that eutectic E_1 (Fig. 1.8) with the composition 66.5% WO_3 and a melting point of 930 (970)°C forms between $PbWO_4$ and WO_3.

There are two compounds in the system, $PbWO_4$ and Pb_2WO_5. The melting points of the two eutectics E_2 and E_3 are 893 (915)°C and 720 (730)°C. The relevant compound – lead tungstate ($PbWO_4$) – melts congruently at 1123°C.

The equilibrium diagram makes it possible to determine the most frequent requirements on the technology of production of the PWO crystals by the Czochralski method. Firstly, because of the high melting point it is necessary to use platinum or iridium crucibles. Secondly, the fact that it is possible to work with both the stoichiometric mixture of the oxides of tungsten and lead and with the excess of either WO_3 or PbO, if this is dictated by the requirements on the quality of crystals, for example optical crystals. Thirdly, the size of the growing crystals is restricted only by the technological possibilities of equipment and the crucible dimensions. Fourthly, there is no urgent need for creating considerable supercooling at the solidification front.

Taking these basic requirements, a technology was developed for the mass production of fast-acting and radiation hard PWO crystals.

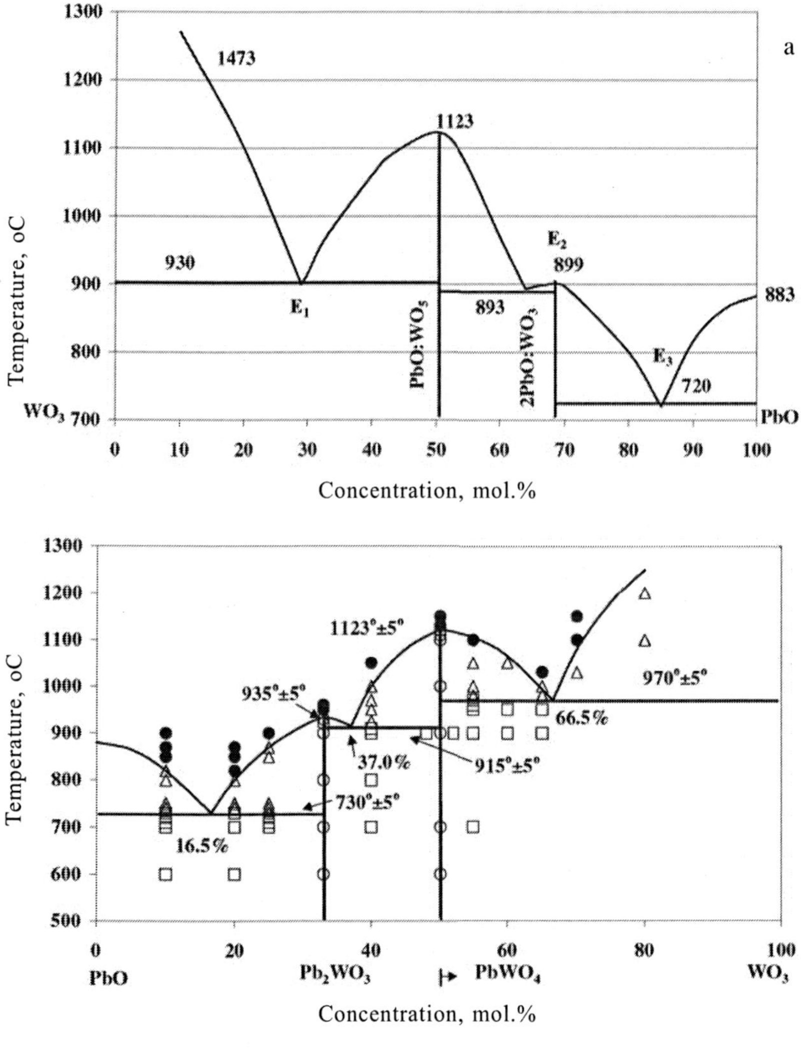

Fig. 1.8. Equilibrium diagram of the PbO–WO_3 system (a – [27], b – [28]).

1.3.3. Point defects and scintillation properties of PWO

The real crystals of lead tungstate, like any other crystals, contain various defects. The industrial production of PWO crystals is characterised by a relatively high rate of mass formation, i.e., in the conditions far away from equilibrium. Therefore, these crystals are not ideal and contain various defects, including point defects. The detailed study of the entire range of the defects is not the aim of this book and attention will be given mainly to point defects. These defects form a set of scintillation properties and, consequently, are the key problem of the technology of production of crystals with the required properties.

The point defects, forming additional levels in the forbidden band, may take part in the formation of optical centres, fulfilling various functions:

- the point defects, trapping and releasing the excited states (excitons, electrons, holes), can play the role of luminescence centres. The nature of induced luminescence depends on the lifetime of the excitation on the level of the defects: in the case of short lifetimes, this may be scintillation, and with increasing lifetime the luminescence is regarded as phosphorescence;
- the capture of excitation on the levels of the point defects may lead to radiationless transitions with the transfer of excitation energy to the phonon spectrum.

If a crystal containing point defects is used as a high-speed scintillator (and the conditions of application of PWO in ECAL CMS are exactly like this), then any point defect with the lifetime greater than 10^{-9} s is regarded as a trap of excited states and leads to extinguishing of scintillations.

Evidently, because of the relatively low yield of the scintillations in the PWO crystals at room temperature, any physical process leading to additional extinguishing of the scintillations should be minimised. In particular, the concentration of the electron traps capturing of the electrons and causing a reduction in the rate of scintillation or the redistribution of the stored light sum to phosphorescence should be reduced. This is why it is very important to know the mechanisms of formation and controlling the point defects in PWO.

To explain the nature of electron and hole centres on the basis of the intrinsic point defects in the PWO crystals, experiments have been carried out using several methods. Special attention was given to the structural special features of the lead tungstate crystals, measurements were taken of the electronic paramagnetic resonance (EPR) in a wide temperature range [29–31], and data have been obtained on the thermally stimulated luminescence (TSL) of the crystals, grown in different conditions [32–34], thermally stimulated conductivity (TSC), photostimulated EPR and the optically detected magnetic resonance (ODMR) [35]. The detailed analysis of the results has made it possible to draw several conclusions regarding defect formation in the PWO crystals.

The structure and properties of the PWO depend very strongly on the growth conditions. The growth of PWO crystals by the Czochralski method from a stoichiometric melt is characterised by a shortage of lead leading to the formation of cation vacancies V_c in the position of the lead ions in the crystal lattice. Analysis of samples of the material evaporated from the crucible during the crystal growth process and settled on the wall of the cooled chamber shows that it may be characterised by the structural formula $Pb_xW_yO_z$, where the value of y does not exceed 0.8 of the value x. In a number of cases and, in particular, in the case of the increase of the number of consecutive growth processes from the same crucible, the value of y decreases even further. The systematic shortage of lead in the crystal formed in the process of synthesis and growth results in additional special features of the crystal structure. These special features are determined by combined measurements using the methods of x-ray and neutron diffraction [23]. The experimental results show that the structure of the crystal remains the same and consists of the regular and automating packing of Pb and W atoms in every column along the axis z in both pure PWO and in the final complicated composition $Pb_{7.5}W_8O_{32}$. In this composition, the parameters of the elementary cell are $a = b = 7.719$ (2), $c =$ 12.018(2) Å with the spatial group $P\overline{4}$.

It can be assumed that the systematic shortage of lead in the crystal is accompanied by the formation of a superstructure of cation vacancies which is compensated by shifts of the atoms as a result of a charge shortage.

This has also been confirmed for the crystals produced by the modified Bridgman method [26, 36, 37]. At a systematic shortage of lead in the crystal, this phenomenon results in the formation of a superstructure [23] due to the ordering of defects based on cation vacancies.

Electron or hole centres with the paramagnetic ground state have not been detected in PWO crystals. This means that the electronic centres of the type F^+ (anion vacancy V_0 + e) or hole centres O^- have no localised energy levels in the forbidden band. Therefore, the cation vacancies, trapping two holes of the type O^-V_cO and oxygen vacancies, trapping an even number of electrons, are most likely to form in the PWO crystals.

In the non-doped PWO crystals irradiated with ionising radiation or with ultraviolet radiation with the photon energy higher than the energy of the forbidden band investigations by EPR at a reduced temperature have not shown the formation of characteristic hole centres $(WO_4)_2^{3-}$. However, several other types of paramagnetic centres were found.

All the PWO crystals are characterised by the presence of a polaron centre – an additional electron stabilised by the Jahn–Teller effect in the regular oxy-anion complex WO_4^{3-} [29, 38, 39]. This is the first of the open characteristic centres. According to the EPR data, the electronic centre WO_4^{3-} is thermally activated in the vicinity of 50 K and its thermal activation energy is E_{TA} = 50 MeV. The WO_4^{3-} centre is dominant at low temperatures (T < 50 K). During thermal activation of this centre, the released electrons partially recombine by radiation and are partially trapped by deep traps.

The second detected characteristic centre is $Pb^{1+} - V_0$ which is stable in a crystal up to 175 K [40, 41]. Substituting the lead iron, the other ion can form the same centre in the vicinity of the anion vacancies, and the essential condition is that the electron is captured in a trap by a heterovalent cation in the vicinity of the oxygen vacancy. Photoionisation of this centre by infrared radiation with a threshold of E_{PI} = 0.9 eV was observed. If irradiation with light takes place at T < 50 K, polaron centres are recreated. The authors of [31] observed a correlation of the thermally initiated breakdown of the centre and a band in the TSL with a maximum at 186 K. It has been shown that thermal activation of this centre is characterised by the energy E_{TA} = 0.55 eV and its thermal activation is accompanied by green luminescence of WO_3 groups. Since $E_{TA} < E_{PI}$, the conduction band does not take any part in the final thermal activation stage and delocalisation of the electron takes place in the direction to the oxygen vacancy with subsequent recombination. During thermal activation of this centre the released electrons are trapped by other electrons with similar energy in the forbidden band or by deeper traps.

Impurity centres also form electronic capture centres. The centres formed during the isomorphous substitution of the tungsten ions and by the molybdenum ions have been described in detail in [42]. Substitution of the lead ions is characterised by two typical cases:

- the localisation of ions with the stable valence state, such as lanthanum, lutecium or yttrium in the position of the lead ions results in the formation of the WO_4^{3-} electronic centre which forms on the basis of regular tungsten anion complexes disrupted by the adjacent rare earth trivalent ion of the impurity [30, 38]. The centre is thermally activated at 97 K and has E_{TA} = 200 meV.

- localisation of the ions with a variable valence state, such as the copper ion Cu^{1+}, in the position of the lead ions results in the formation of electronic centres when the electron, like WO_4^{3-} is stabilised on the tungsten ion or in its vicinity.

It should be mentioned that the 'pure' WO_4^{3-} centre can be stabilised by different methods. Figure 1.9 shows spectra of thermally simulated luminescence in pure crystals, grown in different conditions and activated with yttrium. It may be seen that all the crystals contain a TSL band of approximately the same intensity with a maximum in the vicinity of 50 K, and the appearance of the band does not depend on the growth atmosphere nor the presence of an yttrium-type impurity. This band is associated with the finest of the currently identified electronic centres – WO_4^{3-} (W^{5+}).

This centre can also be stabilised at room temperature but only in the vicinity of a point or substitutional defect. According to the authors of [43, 44], the arguments in favour of stabilisation of this defect simultaneously by an oxygen vacancy and the impurity centre are: the results of annealing in an oxygen atmosphere in which the centres were initially completely annealed and then were not reduced in repeated annealing in argon, and also the absence of these centres in the crystals activated with lanthanum. The authors did not take any account of the simultaneous and obvious conversion of the Cu^{1+} to Cu^{2+} ions in oxygen annealing, accompanied by annihilation of W^{5+} centres.

The presence of the copper impurity in the investigated crystal with the concentration of several ppm was also confirmed by the authors (Annenkov, et al.) using the GDMS method. After repeated annealing in argon there were no copper ions in the single-valence condition and no reduction of W^{5+} was observed.

At the same time, the previously described TSL spectra show that the formation of the 50 K band which is associated in all likelihood with W^{5+} does not depend on the type Y impurity (and, consequently, La) nor on the growth atmosphere. This means that there are two types of W^{5+} centres in the crystal, those which are stabilised at $T < 50$ K on non-perturbed regular oxyanion complexes, and those which also stabilise on the regular oxyanion complexes but are perturbed by a defect.

According to [45, 46], in addition to the distortion of the regular oxyanion complexes, the La impurity ions can form complex associates with cation vacancies [$2La^{3+}$–V_{Pb}]. In this case, two lanthanum ions, substituting the lead anions, are found in the vicinity of the cation vacancy.

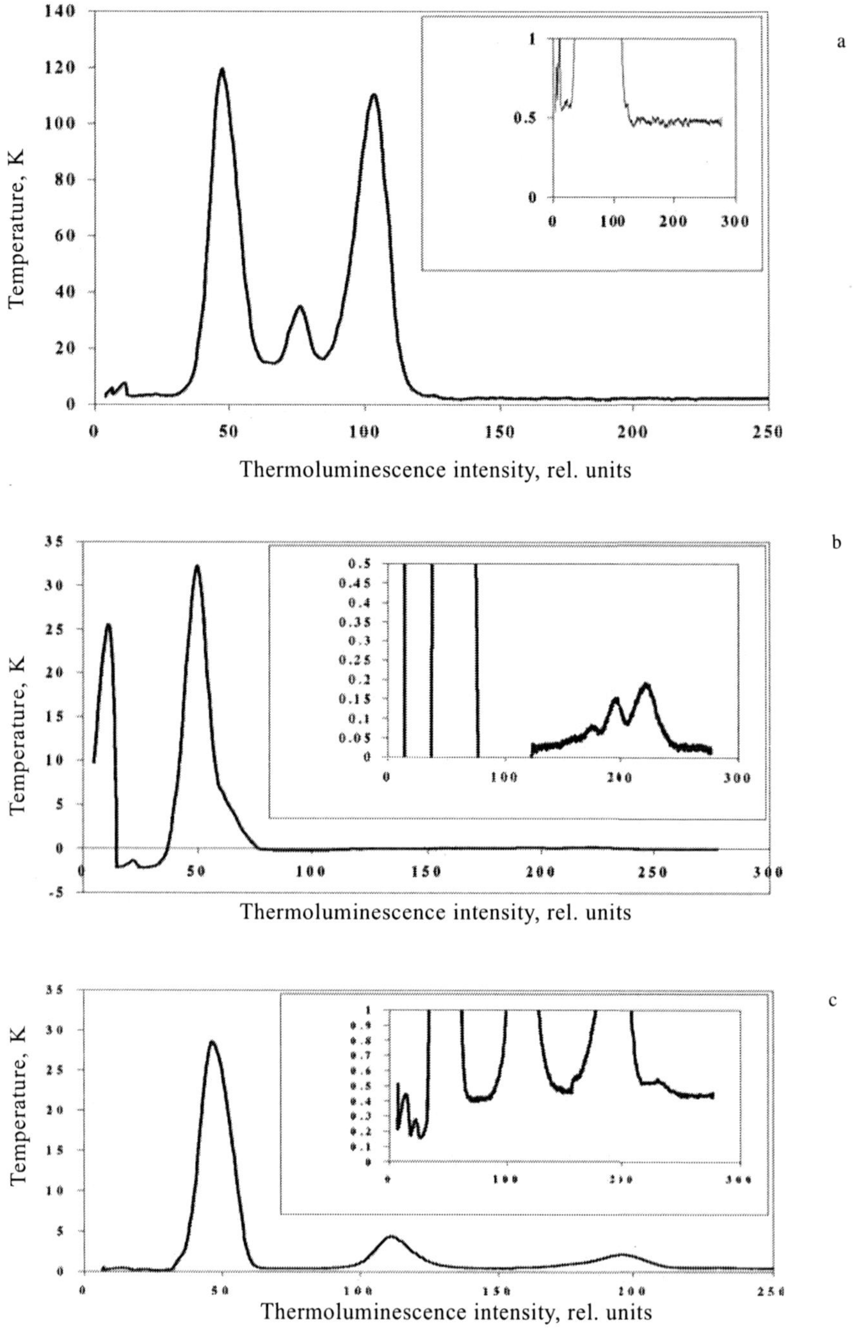

Fig. 1.9. Spectra of thermally simulated luminescence measured in the integral regime: a) non-doped crystal, grown in nitrogen, b) non-doped crystal, grown in oxidation atmosphere, c) crystal doped with Y and grown in oxygen.

Thus, the PWO crystals contain four groups of TSL peaks [47], localised in the following temperature ranges:

I	–	50 K
II	–	70–120 K
III	–	180–240 K
IV	–	300–450 K

The intensity of the peaks of the first group is only slightly sensitive to the crystal production conditions and is associated with auto-localised excitons in oxyanions $[W^{5+}O_4]^{3-}$.

In the groups II–IV the intensity of the peaks depends strongly on the composition of the gas atmosphere in growth, the presence of the impurity centres and the deviation of the crystal from the stoichiometric composition. These peaks are suppressed in the growth of crystals in an oxygen-containing atmosphere or in doping with ions of La, Y and other ions, reducing the concentration of the oxygen vacancies. Consequently, the nature of the peaks is probably linked with the vacancy defects. It is important to note that the peaks of the groups II–IV are characterised by the lifetimes from 10^{-6} to 10^{-2} s, i.e. they are traps and cause luminescence decay. Therefore, the working centre generating the required scintillation can only be represented by a centre thermally released already at a temperature of 50 K. The other centres formed in higher temperatures are traps and inhibit the formation of rapid scintillations.

The peak which breaks down at 50 K is also not an elementary peak and is determined by a group of centres some of which are subjected to the effect of adjacent defects [32]. This is indicated by the shift (Fig. 1.10) of the TSL in the direction of longer waves in multiple successive crystallisation from the same crucible, i.e., in buildup of defects in crystals.

The buildup of defects in the crystals in successive crystallisation processes is indicated by the intensification of the TSL at $T > 100$ K (Fig. 1.11).

The peaks thermally activated at temperatures 330 and 400 K above room temperature correspond to the electronic centres with E_{TA} = 580 and 700 meV, respectively. It should be mentioned that the calculated decay time of the electronic centre with E_{TA} = 580 meV at room temperature is in good agreement with experimental data [48-52]. On the other hand, the decay time of the electronic centre with E_{TA} = 700 meV at room temperature is considerably shorter than that presented in [53], possibly as a result of the insufficient accuracy of the approximation and the low intensity of this peak. In thermal activation both peaks show only red luminescence, are annealed in

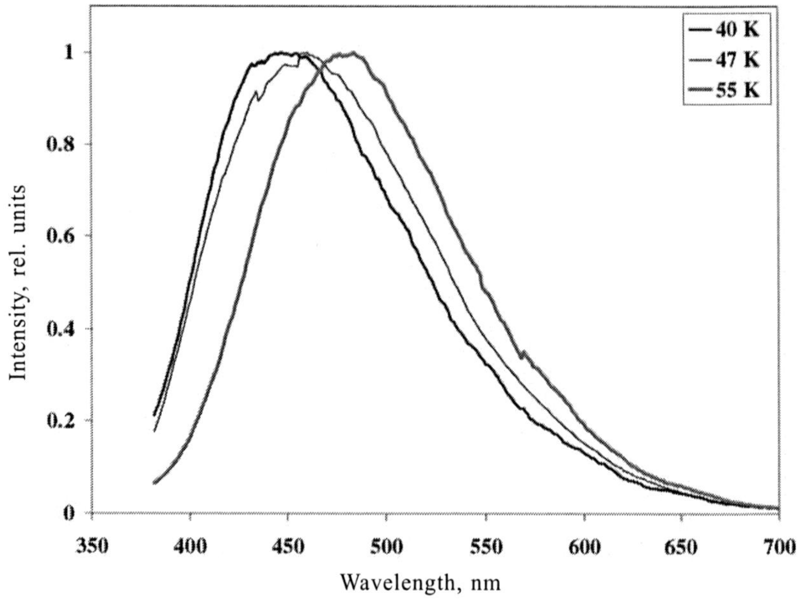

Fig. 1.10. The luminescence spectra in thermal activation of a peak in the vicinity of 50 K in its low-temperature part (40 K), in the vicinity of the centre (47 K) and in the high-temperature part (55 K).

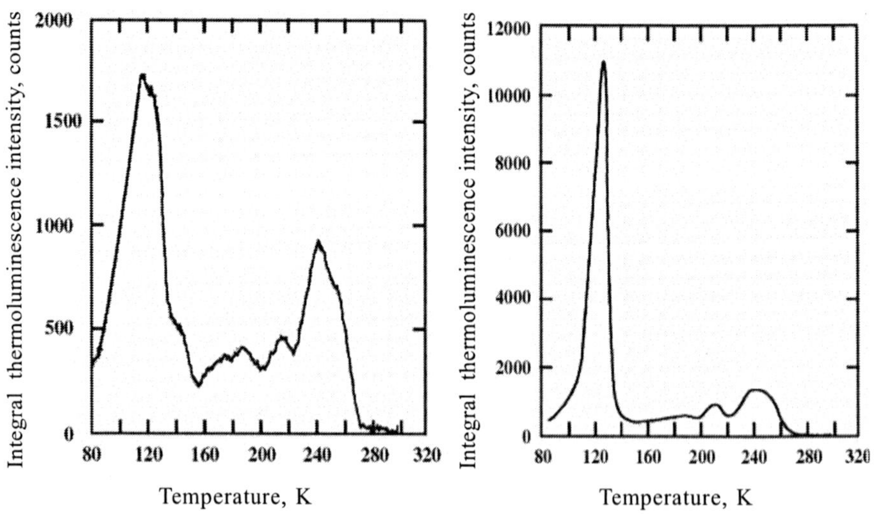

Fig. 1.11. Integral spectrum of thermally simulated luminescence of non-activated crystals, produced after the first (left) and ninth (right) successive crystallisation from the same crucible in air.

the oxidation atmosphere and are suppressed by the impurity centres Y and La, i.e., they are also associated with the presence of oxygen vacancies in the crystal. The centre with E_{TA} = 580 meV is connected with the localisation of two electrons on the oxygen vacancies found at the adjacent tetrahedrons $(WO_3–WO_3)^{2-}$. This relationship requires additional confirmation because the only argument in its favour is the very fact of stability of these associates. This type of double vacancy appears after neutron irradiation and is stable at room temperature, for example, in $CaWO_4$ crystals [54].

As already mentioned, in irradiation of the crystal with ionising radiation even small isolated hole centres situated in the vicinity of cation vacancies are not found in the lead tungstate crystals in a wide temperature range. In addition, doping with pentavalent niobium ions which substitute the tungsten ions in the lattice does not lead to their localisation in the crystal lattice. Both these special features can be explained on the basis of the analysis of the electronic band structure of the crystal. As a result of the localisation of the 6*s* state of the Pb^{2+} ions below the valence band, both centres, V_cO^- and $Nb^{5+}O^-$, have virtual levels inside the valence band, i.e., are delocalised. Only $O^-V_cO^-$ is metastable, resulting in the formation, in the irradiated PWO crystals, of a wide band of absorption induced by ionising radiation with a maximum in the vicinity of 620 nm. However, this centre is also fine since the induced band at 620 nm completely disappears in annealing at 480 K. The approximate energy of thermal activation of $O^-V_cO^-$ is 0.7 eV.

Figure 1.12 shows the scheme of energy levels of characteristic electronic and hole centres in the forbidden band of a lead tungstate crystal [55].

A number of processes taking place in the crystal confirm the proposed energy scheme:

1. The long wave ($\lambda > 700$ nm) part of the spectrum of radiation-induced absorption becomes transparent in irradiation with infrared radiation [50]. This takes place as a result of the ionisation of deep electronic centres.

2. The ionisation of $Pb^{1+}–V_0$ and the sets of the centres $(V_0–V_0)^{2-}$ result in the formation of a high intensity and wide band of short-life absorption in the crystal in the spectral range at approximately 1000 nm [56, 57].

3. Infrared irradiation at 1 eV results in a highly efficient transformation of some electronic centres into others, as indicated by the dependence (Fig. 1.13) of thc intensity of the EPR signals of the previously described electronics centres WO^{3-}_4, WO^{3-}_4–RE (Y),

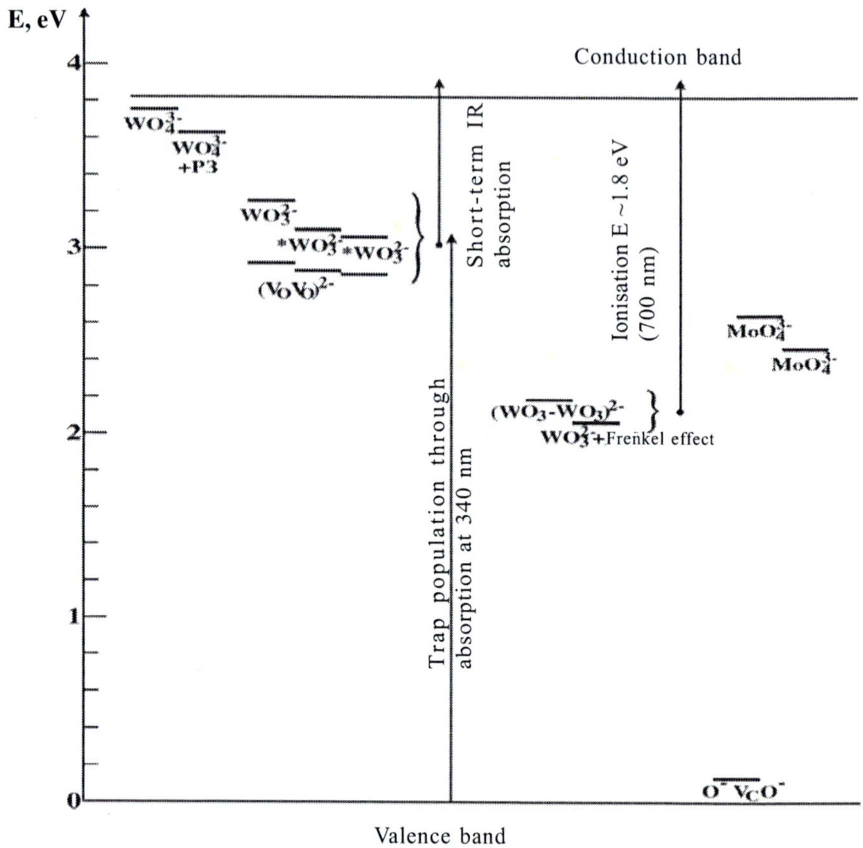

Fig. 1.12 Energy levels of the electron and hole centres in the forbidden band of lead tungstate crystal.

Pb^{1+}–V_0 on the variation of the energy of light quanta in the non-activated crystals. When the energy approaches 1 eV, ionisation of Pb^{1+}–V_0 centres takes place with the ejection of electrons into the conduction band. The formation of polaron centres is accompanied by the formation of axial centres in the crystal based on the non-controlled impurity of Mo ions.

4. The traps interacting with the centres emitting green and red luminescence are populated not only through the conduction band but also during excitation of a set of lower triplet states in WO_3 groups in irradiation of the specimens by ultraviolet light with a wavelength close to 340 nm [58].

The scheme of the energy levels and spontaneous deactivation time of the electron centres at room temperature may be used to estimate the effect of individual types of defects on the scintillation properties

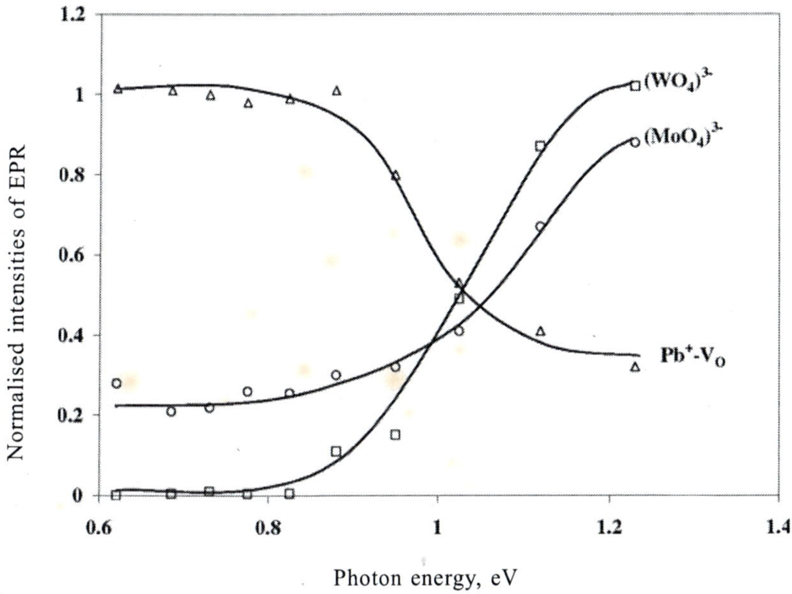

Fig. 1.13. Dependence of the intensity of EPR signals of electronic centres WO_4^{3-}, WO_4^{3-}–Y, Pb^{1+}–V_0 with the variation of the energy of light quanta in non-activated crystals according to the data in [41], T = 15 K.

of the crystals. Only the small traps, forming WO_4^{3-} electronic centres by the relatively rapid transfer of the electrons into the conduction band contribute to scintillation. Others, based on anion vacancies or their associations, provide a contribution to the slow components in scintillation, phosphorescence, and also additional optical absorption, induced by radiation. The hole centres contributed only to optical absorption induced by ionising radiation.

The search for methods of increasing the light yield of the lead tungstate crystals has always been of considerable interest to researchers. For example, the authors of [59] investigated different impurities in lead tungstate crystals in order to redistribute the energy of radiationless losses to the radiation of scintillations from fine electron traps. Figure 1.14 shows the results obtained in activation of a crystal by a specific type of ion.

The highest values were obtained in activation of crystals by Mo and Cd ions. In co-activation with several impurity centres, it has been established that the crystals with the Mo and Nb impurities show at room temperature a scintillation yield of 58 photoelectrons/MeV, those with the Cd and Sb impurity 44 photoelectrons/MeV, and those with the impurities Mo and Sb 38 photoelectrons/MeV. In triple

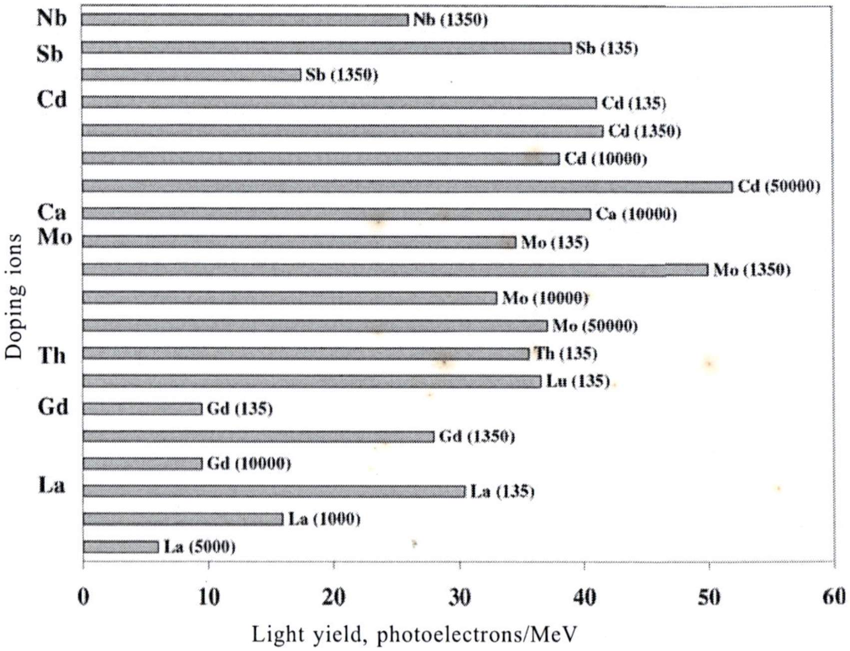

Fig. 1.14. The scintillation yield of PWO crystals activated with a single type of impurity ion (the concentration of the activator is given in the brackets, ppm), T = 300 K.

doping with the ions of Mo, Cd and Sb, the scintillation yield was 75 photoelectrons/MeV.

It should be mentioned that the doping of the PWO crystals with the Mo ions results in the displacement of scintillations into the green region of the spectrum characterised by long luminescence times. On the whole, the attempts to increase greatly the scintillation yield in the lead tungstate crystals by the redistribution of the energy of radiationless recombination to the radiation of scintillations as a result of formation of fine electron levels in the crystal have not as yet been successful.

In addition to the experiments with the activation of the crystal with the impurity centres, the authors of [60] investigated the possibility of redistributing the electronic density of the states in the vicinity of the bottom of the conduction band by changing the composition of ligands in the crystal. The Bridgman method was used to grow lead tungstate crystals with the addition of lead difluoride PbF_2. The activator was used as a super-stoichiometric addition to the stoichiometric composition. The authors recorded an increase of the scintillation light yield as a result of the redistribution of luminescence energy with a clearly increased contribution of the nonregular centres WO_3. The radio-luminescence spectrum of these crystals is characterised by the dominance of the band in the vicinity of 490 nm, and slow components are detected in the scintillation kinetics at room temperature. According to the authors, the fluorine ions partially substitute the oxygen ions, redistributing at the same time the electronic density of the top of the conduction band. The localisation of the fluorine ions F^- in the interstitial space may be regarded as an alternative hypothesis. In this case, the compensation of the surplus negative charge is compensated by oxygen vacancies, i.e., the crystal contains a surplus concentration of WO_3 and, consequently, green luminescence dominates in the crystal. The role of green luminescence has already been discussed.

1.3.4. Special features of using PWO crystals in the electromagnetic calorimeter of the CMS experiments

The main factors for using the PWO crystals in the electromagnetic calorimeter of the CMS experiments at the CERN are: light yield, operating speed and radiation hardness.

Since the light yield in the lead tungstate crystals depends, in addition to other factors, also on the region of the spectrum in which luminescence takes place, this imposes stringent restrictions and requirements on crystal growth technology. The PWO contains two main spectral luminescence regions: shortwave ('blue') band in the range 390–420 nm, and the group of long wave ('green') bands in the range 480–520 nm. According to the current views, the 'blue' luminescence band in the PWO crystal is caused by intra-centre transitions in isolated groups $(WO_4)^{2-}$. 'Green' radiation is caused by the presence of WO_3 groups representing the oxygen tetrahedrons around the W^{6+} ion which contains the oxygen vacancy $[(WO_4)^{2-}–V_O]$ or centres of the type $[(WO_4)^{2-}–F]$.

The light yield of the 'green' PWO crystals is higher than that of the 'blue' crystals. On the other hand, the operating speed of the former is lower. Both these parameters are very critical for the calorimeter. From the viewpoint of technology, the control and fine regulation of the structure of the crystal defects is a very complex task. The reproducibility of technology in mass production is associated with problems and is at least economically inefficient. A compromise is a solution in which the 'blue' crystal is used for the calorimeter and the problem of the light yield is compensated to a large degree using avalanche light diodes with a large sensitive surface.

Figure 1.15 shows a section of the electromagnetic calorimeter based on PWO crystals and a PWO crystal with the installed photodetector.

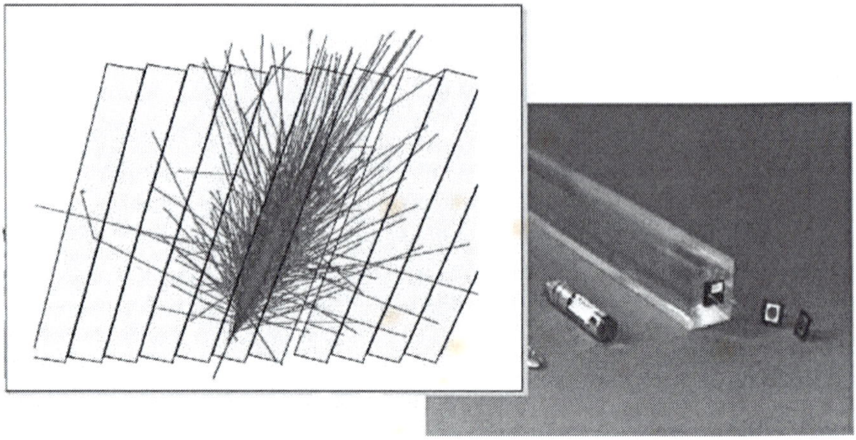

Fig. 1.15. Section of ECAL CMS based on PWO crystals [61].

It should also be mentioned that the temperature dependence of the scintillation yield has a negative coefficient of –2 %/°C in a relatively wide temperature range [62] so that it is possible to increase the scintillation yield further by cooling the detector.

Initial shortcomings of the PWO crystals are the low light yield (no more than 4 photoelectrons/MeV) as a result of high internal optical absorption in the wavelength range of the scintillations, lower radiation hardness manifested in a decrease of the light yield under the effect of the ionising radiation by 20% or more, the presence of slow components in the scintillations, and after glow.

The results of the investigations carried out at the Institute of Nuclear Problems (INP, Minsk, Belarus), the Institute of High-Energy Physics (IHEP, Protvino, Russia) and the Bogoroditsk Techno-Chemical Plant (BTCP, Bogoroditsk, Russia) were used to produce specimens of crystals with satisfactory physical parameters. They are presented in Table 1.3, in comparison with the parameters of some other heavy scintillators and the parameters of the PWO crystals available prior to the start of development of mass production technology. The mass production of the PWO scintillation elements was started in 1998 in Russia at the BTCP (Fig. 1.16).

The result of the investigation and development of mass production technology was the preparation of a specification for the scintillation elements produced from PWO single crystals for ECAL CMS (Fig. 1.17). The most important and critical parameters will now be listed, omitting the relatively strict but standard requirements on the

Table 1.3. Physical properties of PWO crystals in comparison with other heavy scintillators

Property		NaI:Tl	CsI	BGO	YAP:Ce	BaF_2	PWO 1955	PWO 1999
Density, g/cm³		3.67	4.51	7.13	5.35	4.88	8.28	8.28
Refractive index		1.85	1.80	2.15	1.91	1.49/1.57	2.16/2.30	2.16/2.30
Radiation length, cm		2.56	2.43	1.12	2.2	2.03	0.89	0.89
Interaction length, cm		41.4	37.0	21.8	38	29.9	22.4	22.4
Moliere radius, cm		4.80	3.50	2.25	3.7	3.4	2.19	2.19
Spectral luminescence maximum, nm		415	310	505	347	220/310	440–520	440
Relative light yield, %		100	39	19	38	3/23	0.1–0.2*	1.3*
Scintillation luminescence time	τ_1, ns	230	10	60	30	0.6	4 (62%)	4 (95%)
	τ_2, ns			300		620	23 (13%)	15 (5%)
	τ_3, ns						230 (25%)	100 (<1%)

*Element 230 mm long

Fig. 1.16. The first 100 PWO elements supplied to CERN. From left to right: M. Kozhik, V. Kostylev, E. Affray-Hillemans, A. Annenkov, P. Lecoq and I. Tyurin (1998).

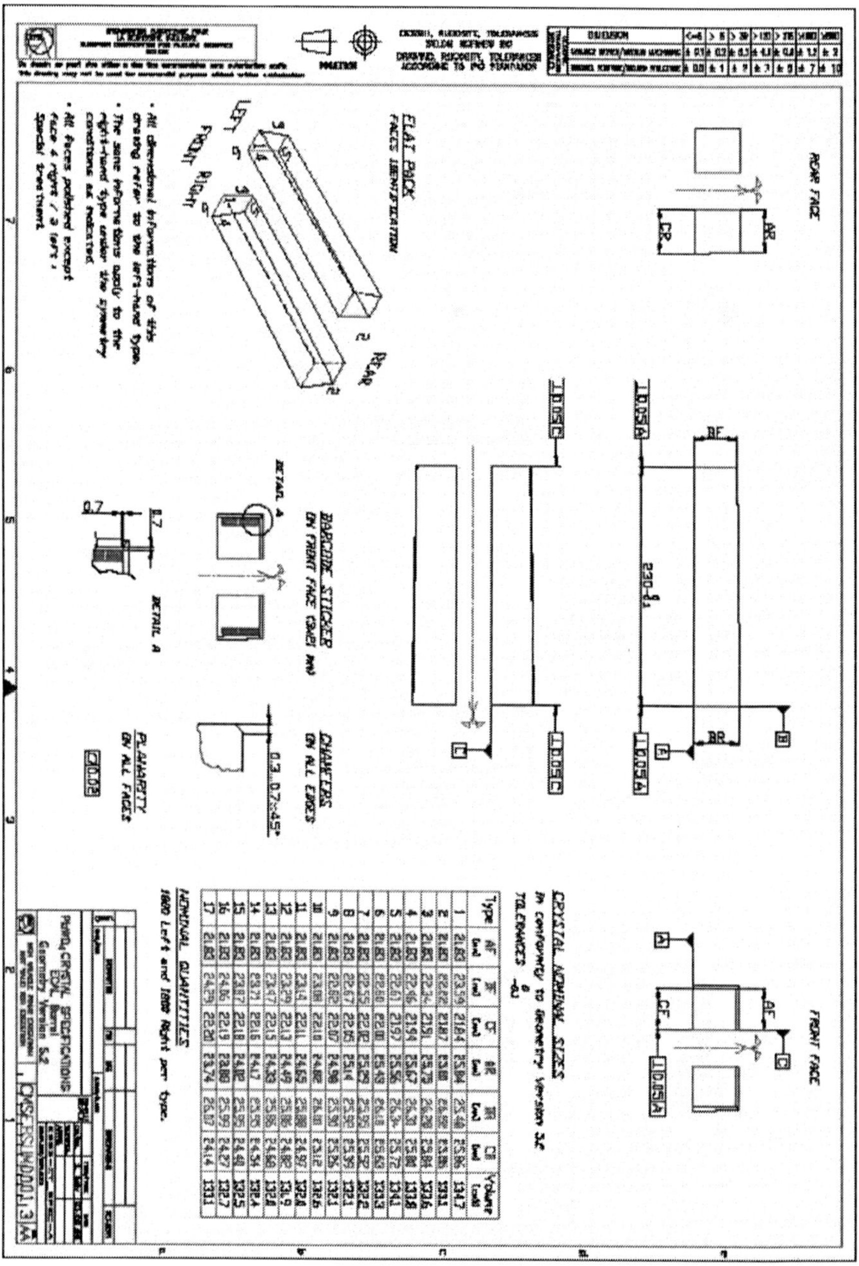

Fig. 1.17. Specifications for the PWO scintillation elements.

dimensions, allowances and the quality of treatment of the surfaces of scintillation elements.

Optical parameters

The transmission spectrum in the direction normal to the crystal axis. This parameter is measured in 10–20 points along the scintillation element 230 mm long in the wavelength range 300–700 nm, with the absolute error of $\pm 0.5\%$. These data are required for verification of the homogeneity of transmission and also represent an additional criterion of the radiation hardness of the crystals.

The transmission spectrum along the crystal axis

This is measured to determine the absorption of scintillations and is used for equalising the light collection in the scintillation elements by additional grinding one of the side faces of the crystal. The absolute transmission values: $T \geq 10\%$ (360 nm); $T \geq 55\%$ (420 nm); $T \geq 65\%$ (600 nm).

The curvature of the edge of the transmission spectrum

This parameter is calculated using the longitudinal transmission spectrum in the wavelength range from 340 to 380 nm. It should be $S > 3.0\%/\text{nm}$. Fulfilment of this condition is an additional criterion of the radiation hardness of the crystals.

Scintillation parameters

Light yield

Light yield is measured in more than 10 points along a crystal 230 mm long followed by evaluation of the heterogeneity of its values. Light yield should exceed 8 photoelectrons/MeV in measurements using an XP2262 photoelectronic amplifier where the entire back surface of the crystal is in optical contact with a photocathode (a window 24×24 mm^2) and a Rhodorsil-type optical lubricant with the refractive index $n = 1.5$ is used, with the integration time of 100 ns, temperature of 18°C and the radiation source positioned 80 mm from the output surface of the crystal.

Heterogeneity of the light yield

This is the relative variation of the light yield along the length of the scintillation element. This parameter has a strong effect on the accuracy of determination of particle energy recorded using scintillation detectors working in the total absorption regime. It is expressed in percent of the radiation length unit. The heterogeneity of the light yield, measured for two ranges: from 35 to 115 mm and from 115 to 185 mm from the front end surface of the crystal, should not exceed $\pm 0.35\%/X_0$ for the first range, and from $-1.80\%\ X_0$ to $+0.25\%\ X_0$ for the second range.

Scintillation kinetics

Scintillation kinetics should satisfy the following criterion: the ratio of the light yield measured for the signal integration time of 100 and 1000 ns should be greater than 90%.

Afterglow

Afterglow should not be greater than 0.5% of the amplitude of the peak in irradiation with γ-quanta with the energy from a Co^{60} source at a counting rate of 1 MHz.

Radiation hardness

Induced absorption

This parameter was measured at a wavelength of 420 nm in lateral irradiation from a Co^{60} source at a power of the absorbed dose greater than 100 krad/h and the absorbed dose greater than 53 krad. Its value should be $\mu < 1.5\ m^{-1}$ at the start of saturation.

Light yield losses

Light yield losses should be smaller than 6% in frontal irradiation using a Co^{60} source at a power of the absorbed dose greater than 15 rad/h and the absorbed dose greater than 200 rad.

An additional criterion is the absence of restoration time constants shorter than 1 h.

2. Control of defect formation in lead tungstate crystals

2.1. Equipment and experimental methods

The individual characteristics of the PWO crystals, both specially grown in experiments and mass production crystals, have been analysed to investigate the processes of defect formation and methods of controlling defectiveness.

Detailed investigations of the microelemental composition of the crystals were carried out using glow discharge mass spectroscopy (GDMS) by Shiva Technologies Europe (France).

The methods of investigating the optical and scintillation properties of the scintillation elements of lead tungstate will be described in detail in section 4. Here, they are mentioned only briefly.

Measurements of light yield and of the nonuniformity of the light field were taken using the method of amplitude analysis on the basis of the position of the total absorption peak in the amplitude spectrum of the ^{60}Co isotope, and also by integration of the time spectra of the scintillations.

The scintillation kinetics were investigated by the start-stop method. The optical transmission of the lead tungstate crystals at different wavelengths was measured using spectrophotometers.

Investigations of the radiation hardness of the PWO crystals were carried out at the Main Radiation Centre (GIF, CERN), on a cobalt source at the Geneva Hospital and at the INP Radiation Centre (Minsk). The main characteristics of these centres may be briefly characterised as follows:

– the Main Radiation Centre has at its disposal the source of γ-radiation ^{137}Cs. The source ensures lateral irradiation at the power of the absorbed dose of 15 rad/h, which is close to the actual conditions of irradiation inside the working CMS detector. In the Centre, the radiation damage and also the kinctics of radiation damage were controlled by measuring the losses of optical transmission of the full-size crystals. It was also possible to carry out periodic irradiation with an electron beam in order to take direct measurements of the change of the scintillation signal during irradiation. A shortcoming of this centre is the extremely low productivity;

– the Radiation Centre at Minsk uses the source of γ-radiation ^{60}Co generating radiation at a power of the absorbed dose of the order of 10 000 rad/min. In this centre, radiation damage is controlled by measuring the losses of optical transmission of small crystal specimens at the saturation dose. The upper parts of crystal boules (the seed region of the crystal) were investigated here;

– the centre at the Geneva Hospital uses a source of ^{60}Co providing radiation and a power of the absorbed dose of the order of 25 000 rad/h. Equipment makes it possible to control the radiation damage by measuring the induced optical absorption of full-size crystals at the saturation dose.

The semiautomatic RGB system is fitted with a source of ^{60}Co generating radiation at a power of the absorbed dose of the order of 10 000 rad/h. Thus, RGB can be used to control both radiation damage and the kinetics of restoration of radiation damage by measuring the induced optical absorption of the full-size crystals at the saturation dose. The system is described in more detail in paragraph 4.2.3.

The experimental results were processed by the methods of mathematical statistics and correlation analysis.

2.2. Radiation hardness and formation of point defects in PWO crystals

The radiation hardness of the PWO crystals for application in ECAL CMS is regarded as the variation of the optical properties under the effect of gamma irradiation from ^{60}Co sources. The effect of radiation is determined by the change of the state (charge exchange) of the intrinsic point defects present in the crystals which form during growth and heat treatment, and defects formed directly under the effect of radiation. The point defects, both intrinsic and those formed under the effect of irradiation, influence the scintillation characteristics on the one hand and the formation of colour centres on the other hand.

The possibilities of increasing the structural perfection will now be discussed. One of the possibilities of producing point structural defects in the crystals, investigated in paragraph 1.2.3, is doping of the crystal with the ions of Y and La compensating the shortage of Pb^{2+} in the crystal and suppressing the formation of oxygen vacancies. In fact, at low concentrations of the lanthanum ions (< 50 ppm) the scintillation yield of the crystal increases by 30–50%. At the same time, an increase of the concentration of these activators results in recapturing of the electrons from polaron centres, preferential capture of the electrons by traps based on distorted regular oxyanion complexes of rare-earth metals, and also localisation of interstitial oxygen ions in the crystal.

Previously, it was mentioned that the decrease of the defect concentration by means of doping does not improve the operating speed of the PWO scintillators and displaces the radiation spectrum into the long wavelength range.

Another method of reducing the concentration of point defects may be the combination of low doping (impurity concentration not greater than 50 ppm) and strict control of the stoichiometry of the melt from which the crystal is grown. Combined application of doping and strict control of the melt stoichiometry remove both the systematic and fluctuation reasons for the formation of vacancies in the crystal.

Specimens of crystals with the volume of several tens of cubic centimetres, grown in a nitrogen atmosphere, with the lanthanum concentration than 40 ppm, have the light yield of 45 photoelectrons/MeV. The full-size crystals 23 cm long produced from the same crystal have the light yield not lower than 20 photoelectrons/MeV at 18°C.

Figure 2.1 shows the time dependence of the integral of the scintillation pulse of the $PbWO_4$:La crystal measured at different temperatures. At a temperature slightly lower than 0°C the light yield of 80 photoelectrons/MeV can be reached in these crystals, 95% of the light of the scintillation pulse is released during no more than 100 ns.

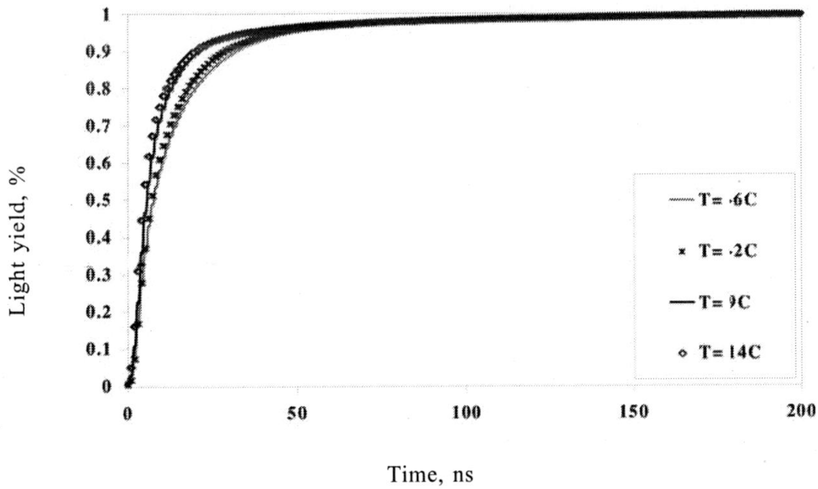

Fig. 2.1. Time dependence of the integral of the scintillation pulse in PWO crystal, measured at different temperatures. The crystal was grown in nitrogen with the lanthanide concentration in the crystal of 20 ppm.

Another method of increasing the light yield of the PWO crystals is based on adding adding additional luminescence centres with specific parameters to the crystal. In this case, an obvious requirement on such a luminescent centre is the effective capture cross-section of the electrons from the conduction band, with the electrons formed as a result of thermally induced dissociation of excitons. Another important factor is the recapture of electrons from fine traps based on oxygen vacancies, i.e., these centres should have the energy of excited states covering the range 3–4 eV and the emitting state smaller than 3 eV. For this reason, the Ce^{3+} ions are not a suitable activating addition. Activators may include the ions of Mo which replace the tungsten ions in the lattice, and also Tb ions which replace the lead ions in the crystal. This has been confirmed by experiments investigating the crystals with different combinations of annealing procedures and additions of Mo, La, Tb, Y [63–66].

The stability of the parameters of the scintillation material in the conditions of high radiation load is another important parameter in application in precision electromagnetic calorimeters. If the depth in which the defects formed in the crystal or defects formed under the effect of radiation are found in the crystal is relatively large in the forbidden band, of the order of 1 eV or greater, their charge exchange in the field of ionising radiation results in changes of the physical properties of the medium, such as optical transmission and electrical conductivity which remaines unchanged for a certain period of time after the end of irradiation.

The charge exchange of the structural point defects and impurity ions results in the observed effect of radiation damage of the crystal even at a moderate absorbed dose [67].

γ-induced radiation damage of a scintillation crystal formed at a relatively low dose power will be discussed. Here no attention is given to the effects of structural changes, i.e., phase transformations of matter and its amorphisation in the conditions of strong radiation fields. If irradiation takes place at room temperature, the effects of condensation of defects and of their absorption by dislocations can also be ignored. In these cases, the recorded radiation damage of the scintillation crystal is reduced to two effects:

1. The change of optical transmission of the crystal which forms as a result of the appearance of point defects and changes in the charge of the existing point defects in the structure and heterovalent random impurities;

2. Radiation damage of the scintillation mechanism as a result of radiation-induced charge exchange of the luminescence centres, causing

scintillation. The dominant component of the damage depends on the mechanism of formation of scintillations, the concentration of relative fraction of the defects and also on special features of formation of the defects in the crystal matrix.

The scintillation mechanism is not disrupted in all crystals. For example, this does not take place in self-activated scintillation crystals $Bi_3Ge_3O_{12}$ where the luminescence centres are represented by matrix-forming oxyanion complexes. Similarly, in lead tungstate crystals there is no damage to the scintillation mechanism under the effect of ionising radiation and there are only changes in the optical transmission of the crystals as a result of the formation of colour centres with absorption bands in the visible range of the spectrum. Following the tradition in CMS collaboration, this phenomenon will be referred to as the radiation damage of PWO crystals, so that it differs from the phenomenon of damage of the scintillation mechanism. With time, the induced absorption spectrum relaxes with constants which depend on the nature of the colour centres and their annihilation mechanisms.

Additional activation with impurity ions is preferred in the group of different methods of simultaneous suppression of both radiation damage components. The increase of the structural perfection of the crystals as a result of the purity of the initial materials and the decrease of the concentration of initial defects also result in a positive effect for self-activated scintillation materials.

Effects of radiation damage of optical transmission in lead tungstate crystals will be examined in greater detail. In the large majority of cases of application of this material, the power of the dose of recorded ionising radiation is such that there are no changes in the structure of the scintillation matrix and there is only buildup of primary defects in the crystal, capture of free carriers by vacancies and their recombination.

Figure 2.2 shows the absorption spectrum, induced by gamma radiation, in a non-doped PWO crystal grown in air from a stoichiometric mixture of oxides and measured 1200 and 6×10^6 s after irradiation. The spectrum consists of a resolved absorption band with the maximum in the vicinity of 400 nm and of a wide non-resolved band extending from 500 to 800 nm. The insert in the same figure shows the change of the normalised values of the absorption coefficient at 380, 440 and 600 nm with time. It may be seen that the decrease of induced absorption at the given wavelengths is governed by a common law.

This means that the recombination of the centres which provide a contribution to the induced absorption in this spectral region is

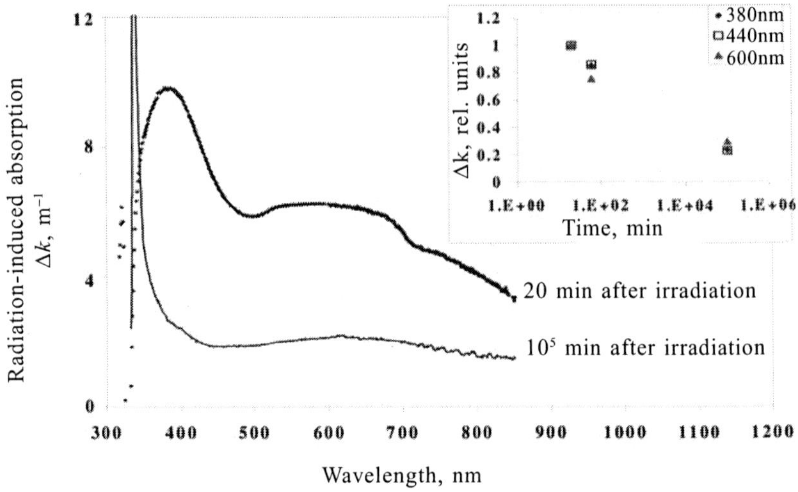

Fig. 2.2. Induced radiation spectra in a non-activated lead tungstate crystal, measured after 1200 and 6×10^6 s, irradiation with a source of ^{60}Co, dose 104 rad/h, radiation time 20 min at T = 300 K.

correlated. As already mentioned, the Frenkel-type defects cause the formation, in the non-irradiated crystals, of an absorption band with the maximum in the vicinity of 350 nm.

Therefore, the loss of the electrons captured by these defects and restoration of the initial concentration are accompanied by a decrease of induced absorption in the spectral range 400–800 nm, which is accompanied by an increase of absorption in the range $\lambda < 380$ nm.

The spectral components of the wide band can be resolved in crystals activated with the ions of La, Y and Nb which greatly reduce the concentration of different types of recharged point defects [68, 69], especially centres based on cation vacancies. In particular, these crystals are characterised by the formation of a wide band of induced absorption with a maximum in the vicinity of 620 nm which masks other bands. This is clearly indicated by the spectra in Fig. 2.3 which shows the induced absorption spectrum of a lead tungstate crystal activated with Y ions with a concentration of ~100 ppm and measured after 1674, 27010, 86185 and 165883 seconds.

The spectrum of this crystal contains three bands with a maximum at 520, 470 and 720 nm. Identical bands are also found in the crystals activated with lanthanum with the concentration of ~20–30 ppm, i.e., in cases in which only part of the vacancies is compensated by the impurity (Fig. 2.4). The crystal also contains both bands 520 and 470nm.

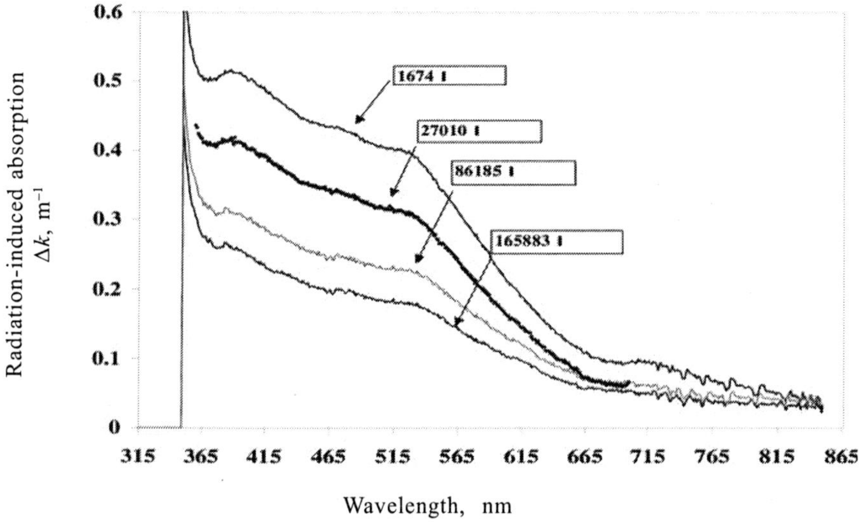

Fig. 2.3. Induced absorption spectra in PWO crystal, activated with yttrium ions with the concentration of 80–100 ppm, measured after 1674, 27010, 86185 and 165883 s, irradiation with a ^{60}Co source, dose 10^4 rad/h, radiation time 20 min, T = 300 K.

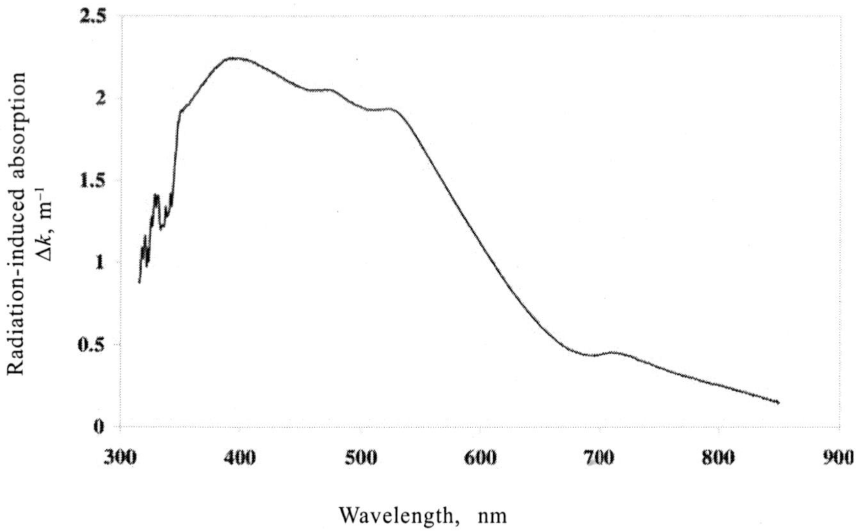

Fig. 2.4. Induced absorption spectrum of a $PbWO_4$:La crystal, 23 cm long, La concentration 20–30 ppm, irradiation with a ^{60}Co source, dose 10^4 rad/s, radiation time 20 min, T = 300 K.

Figure 2.5 shows the induced absorption spectrum of a standard crystal produced for a CMS electromagnetic calorimeter for LHC experiments. The 600 and 720 nm bands are completely suppressed in the spectrum and only weak bands at 400, 470 and 520 nm appear and merge into a single low-intensity wide band.

Attention should be given to the spectrum becoming transparent in the range $\lambda < 400$ nm after irradiation. This effect appears due to the fact that the concentration of Frenkel-type defects in this crystal is low and the deviation from the spectrum of the ideal crystal in this region is also small. Therefore, the effect of radiation conversion of the centres becomes quite evident on the background of edge absorption: the absorption band with the maximum at 350 nm transforms into a band with a maximum in the vicinity of 400 nm.

Thus, the pure and activated PWO crystals contain five absorption bands characterised by relatively slow decay and induced by ionising radiation with the maximum in the vicinity of 400, 470, 520, 620 and 720 nm. According to the previously mentioned classification of the electronic and hole centres, it may be assumed that the 620 nm band is associated with the absorption transition between the ground and first excited state in the $O^-V_cO^-$ centre. This is caused by the fact that the presence of these centres, associated with cation vacancies, becomes

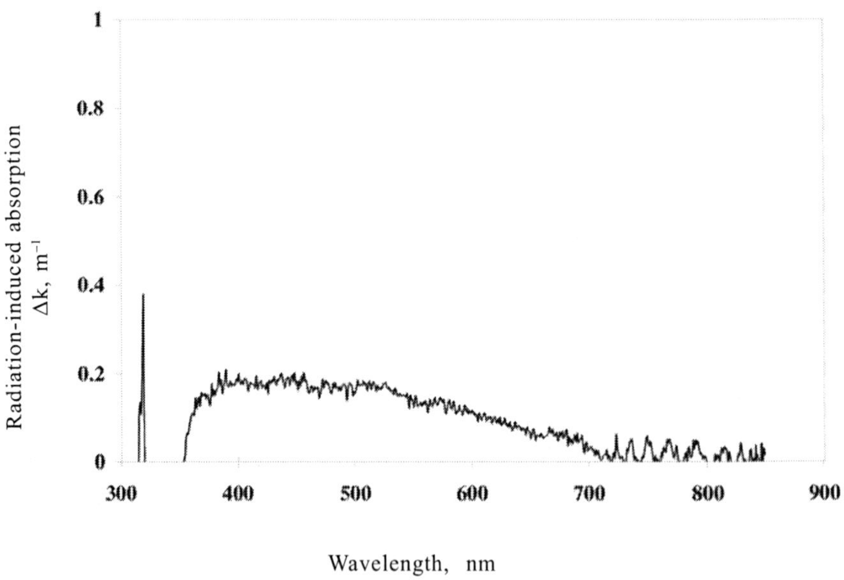

Fig. 2.5. Induced radiation spectrum of a standard PWO crystal, produced for for a CMS electromagnetic calorimeter for the LHC experiments, irradiation with a ^{60}Co source, dose 10^4 rad/h, radiation time 20 min, T = 300 K.

evident in doping of crystals with ions of the type of lanthanum which substituted the lead ions in the lattice.

The 400 nm shortwave band is determined by transition in the anion vacancy in the composition of the Frenkel defect which captured two electrons. The induced absorption band with the longest wavelength of the resolved bands rapidly relaxes and is determined by the intra-centre transition in $(WO_3–WO_3)^{2-}$centres. There are different views on the interpretation of the induced absorption bands at 475 and 520 nm.

The thermal activation time of $(WO_3–WO_3)^{2-}$ centres is approximately 500 s and that of the Frenkel-type defects 10 000 s [47]. These parameters of the kinetics of restoration of the transmission spectra are detected for non-activated lead tungstate crystals. However, their fraction in the kinetics of restoration of optical transmission of the crystals activated with La or Y is very small.

Figure 2.6 shows the relaxation kinetics of induced absorption at 400 and 600 nm in a non-activated crystal. The kinetics were measured 1100 s after irradiation at room temperature. The figure also shows the results of various approximations of the kinetics of decrease of induced absorption. In the exponential approximation corresponding to the decrease of the number of defects only as a result of thermal activation of the centres, the kinetics are satisfactorily approximated by two exponents with constants of 0.5 h and 75 h. If the high-speed component is caused in all likelihood by thermal activation of $(WO_3–WO_3)^{2-}$ centres, then the long component cannot be connected with thermally initiated recombination. It is well-known [47] that the thermal activation energy is in the range from 0.03 to 0.7 eV, and the longest activation time of these centres does not exceed 10^3 s. Therefore, in addition to thermal activation of the centres and subsequently combination of the resultant carriers, in the crystal there should also be other recombination processes caused by, for example, electron tunnelling. In the tunnelling effect, the electron should be realised as a result of its quantum-mechanics nature, with a certain probability of detection, at a relatively large distance from the electronic centre. The electron can penetrate with some probability into the region of the hole centre positioned quite far away (~100 Å) from the electron centre and then recombine with the latter.

Since the concentration of the radiation-induced electronic and hole centres in the lead tungstate crystals is relatively low and the appropriate point defects have no spatial correlation in their localisation in the crystal, the process of recombination as a result of tunnelling can be described by the approximation for randomly distributed interacting centres.

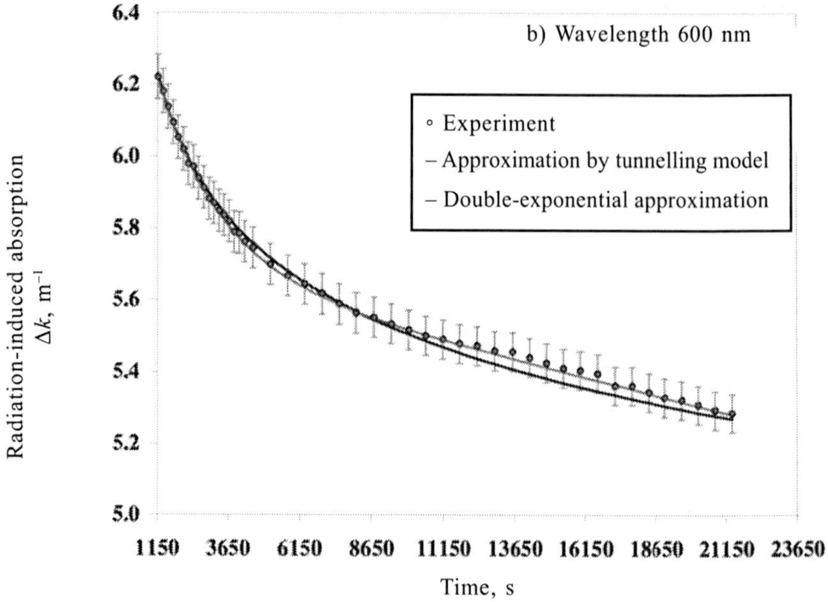

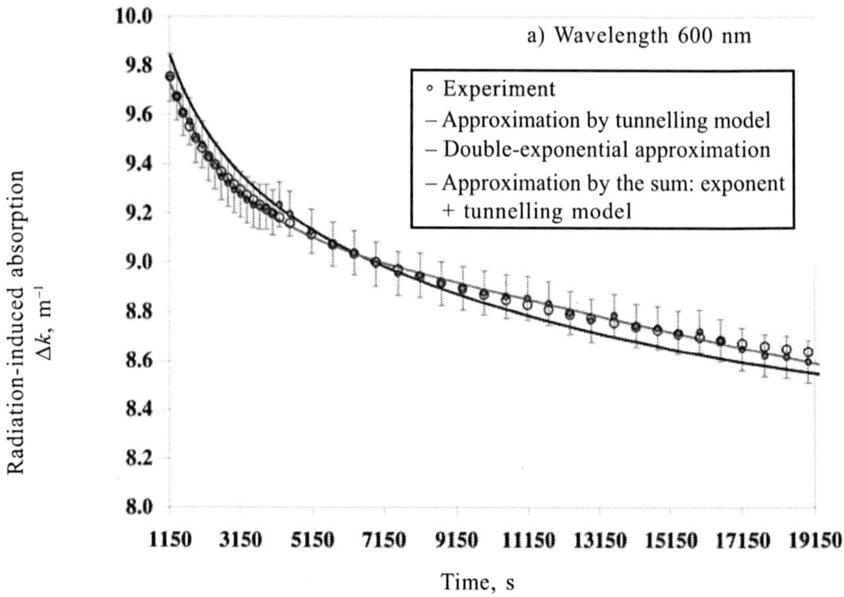

Fig. 2.6. Kinetics of relaxation of induced absorption and different approximations of the kinetics at 400 and 600 nm in a non-activated crystal after irradiation with a ^{60}Co source, dose 10^4 rad/h, irradiation time 20 min at $T = 300$ K.

In addition, a unique situation forms in the lead tungstate in which only two types of centres operate as a result of interaction due to tunnelling, i.e., the deepest defects of the Frenkel type and $O–V_cO^-$ centres.

In addition to approximation by the double-exponential dependence, Fig. 2.6 shows the approximation of the kinetics of relaxation of induced absorption on the basis of the tunnelling model [70].

The same approximation was used for describing the kinetics of relaxation of induced absorption in standard crystals in irradiation with low dose powers. Figure 2.7 shows the relaxation of induced absorption at 470 nm in standard crystals for a CMS calorimeter for the LHC project after irradiation with a ^{60}Co source.

Measurements were taken of the very low values of optical absorption (of the order of 0.1 m^{-1} or smaller) and, consequently, the spectra contain noise and drift of measuring equipment. It may be seen that as in the case of irradiation with a saturating dose, in the case with a small dose the approximation which takes into account the tunnelling effect is in good agreement with the experimental data. The approximation parameters of relaxation kinetics of the induced absorption spectra, shown in Fig. 2.6 and 2.7, with the tunnelling effect taken into account are presented in Table 2.1. In the table, δt is the time between the end of irradiation and the start of measurements, λ is the wavelength at which measurements were taken, τ_i is the time constant of relaxation of i-th exponential component. The sum $\Delta k_{exp}(0)$ + $\Delta k_{tun}(0)$ was assumed to be equal to Δk^{exp}, i.e., the value measured after irradiation after the minimum possible period of time. As already mentioned, in the non-activated crystals the spectral region to which the $(WO_3–WO_3)^{2-}$ centres contribute contains a single high-speed component of relaxation of the induced absorption spectra even after 1100 seconds after irradiation. The high-speed component is close in value to the thermal activation time of the centres at room temperature determined by the TSL method.

An identical component should be detected in the relaxation of hole centres. In fact, the high-speed component of the same order was detected in measurements in the vicinity of 600 nm. At the same time, Table 2.1 shows that the main contribution to the induced absorption amplitude in the non-activated crystals is provided by the centres which are relaxed by means of the tunnelling mechanism.

The best agreement between the approximation results and the experimental results for the non-activated crystals for 395 and 600 nm was recorded at $n(0) = 2.8 \cdot 10^{18}$ and $4.9 \cdot 10^{18}$ cm^{-3}. Since the concentration of Pb^{2+} ions in the crystal is of the order of $11 \cdot 10^{22}$ cm^{-3},

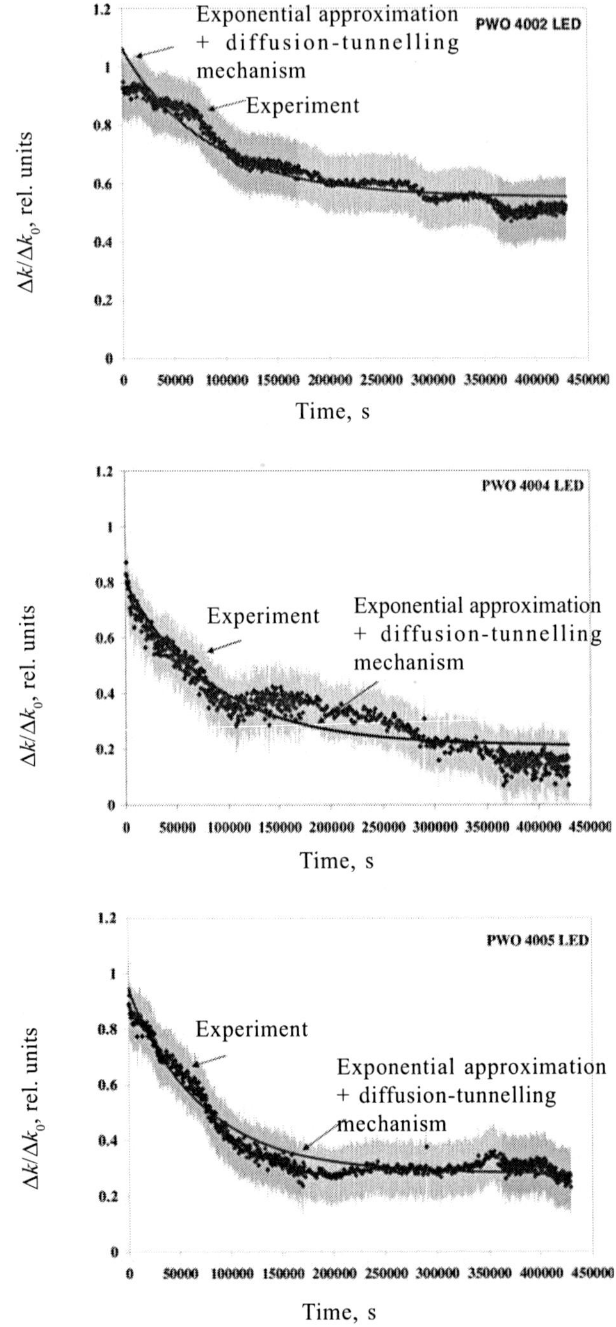

Fig. 2.7. Kinetics of relaxation of induced absorption and approximations at 470 nm standard crystals after irradiation with a ^{60}Co source, dose 15 rad/h, irradiation time 30 h at T = 300 K.

Table 2.1. Approximation parameters of relaxation kinetics of induced absorption in lead tungstate crystals

Crystal	δt (s)	λ (nm)	τ_i (s)	$\Delta k_{exp}(0)$, (cm^{-1})	$n(0)$ (cm^{-3})	$\Delta k_{tun}(0)$ (cm^{-1})
PWO	1100	395	448	1.1	$2.8 \cdot 10^{18}$	20.4
		600	1430	0.2	$4.9 \cdot 10^{18}$	18.7
PWO:Y	1100	400	77000	0.75	$4.1 \cdot 10^{17}$	0.48
		470	75000	0.7	$3.3 \cdot 10^{17}$	0.34
		600	69000	0.36	$2.7 \cdot 10^{17}$	0.16
		700	590	0.4		
PWO #4002	100	470	81500	0.068	$1.7 \cdot 10^{17}$	0.120
PWO#4004	100	470	53170	0.051	$6.7 \cdot 10^{17}$	0.053
PWO#4005	100	470	80000	0.103	$5.6 \cdot 10^{17}$	0.038

the concentration of the defects based on the centres which capture two holes is equal to several tens of ppm. This value is in agreement with the required concentration of the activator, yttrium or lanthanum, for suppressing these defects in the crystal.

In particular, activation of the crystals by these activators is accompanied by the formation of centres on the basis of divacancies and hole centres capturing two holes. Therefore, the relaxation kinetics of induced absorption in the spectral range 400–600 nm does not contain any components with the relaxation constants lower than 1000 s. At the same time, at the 620 nm band the relaxation kinetics of induced absorption at the wavelength of 720 nm can be measured. The calculated constant is 590 s, which is in good agreement with the data for thermal activation of $(WO_3–WO_3)^{2-}$ centres at room temperature. Since the TSL band corresponding to thermal activation of the centres is situated in the vicinity of room temperature, the relaxation kinetics of induced absorption are sensitive to temperature changes in this region. The approximation results also show a decrease of the concentration $n(0)$ of the defects in additional activation with yttrium, as in the case of activation with lanthanum.

It should be mentioned that the proposed method of evaluation of the concentration of the electronic and hole centres capturing two holes and, consequently, of the point defects is based on the measurement of the relaxation kinetics of optical absorption induced by ionising radiation has been used successfully for optimising the concentration of activators to improve the radiation hardness of the lead tungstate crystals.

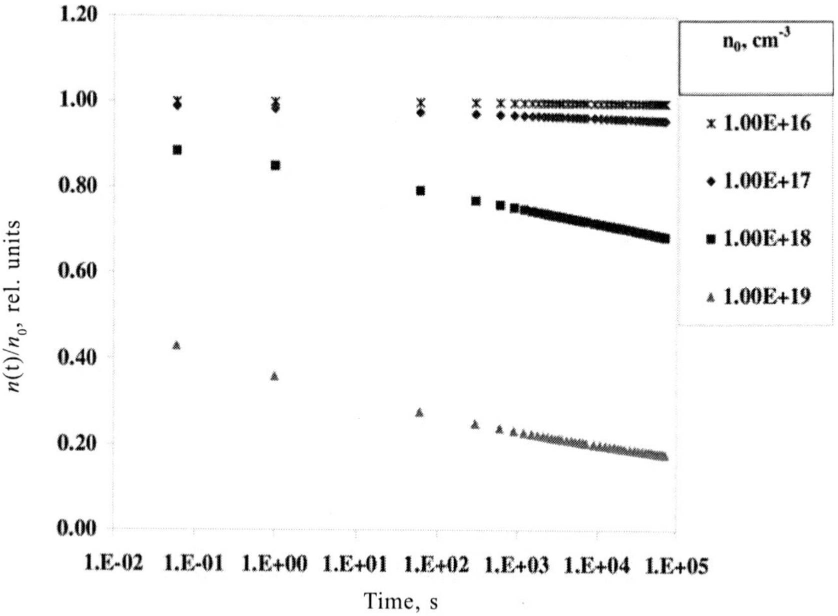

Fig. 2.8. Calculated relaxation kinetics of induced absorption by the tunnelling mechanism in a lead tungstate crystal at $U_{max} - E = 0.7$ eV, $\nu = 500$ cm^{-1}, and the initial concentration of the electronic centres $n(0) = 1 \cdot 10^{16}$, $1 \cdot 10^{17}$, $1 \cdot 10^{18}$, $1 \cdot 10^{19}$ cm^{-3}.

Since the probability of recombination by the tunnelling mechanism depends on the initial concentration of the electronic centres, it is evident that at a higher concentration of these centres tunnelling also contributes to the high-speed relaxation components. Figure 2.8 shows the calculated relaxation kinetics of induced absorption by the tunnelling mechanism in the lead tungstate crystals at various concentrations of electronic centres which loose electrons mainly as a result of tunnelling.

It may be seen that at a concentration of 10^{16}–10^{17} cm^{-3} tunnelling does not contribute to decay in the initial kinetic stage, but at a concentration of ~10^{19} cm^{-3} more than 30% of centres recombine by the tunnelling mechanism in the first 100 seconds after ending radiation.

With the appearance of the high-speed components in the relaxation of induced absorption in the spectral range 400–600 nm, and also with a general decrease of the defect concentration, the contribution of recombination as a result of tunnelling becomes smaller and components with times of $6 \cdot 10^4 - 8 \cdot 10^4$ s appear in the kinetics. These components are of no importance in the relaxation kinetics of the non-activated crystals where the hole centres capturing two holes are dominant on the basis of randomly distributed cation vacancies. The presence of these components correlates with the suppression of the induced absorption band at 620 nm and with the appearance of the 470 and 520 nm bands in the spectrum. Therefore, the formation of these components is caused in all likelihood by the recombination of centres which produce the 470 and 520 nm induced absorption bands.

3. Technology for mass production of lead tungstate single crystals

3.1. Raw material for growing radiation-hard lead tungstate crystals

The specification for the raw material for the production of lead tungstate scintillation crystals by the Czochralski method was prepared on the basis of the correlation of the scintillation parameters of the PWO crystals and the data obtained in analysis of these crystals by the GDMS method carried out by Shiva Technologies S.A. More than 300 specimens were examined. The investigations were conducted on the upper and lower parts of the crystals so that it was possible to calculate the effective distribution coefficient of different ions in the process of crystal growth by the Czochralski method (Table 3.1). Measurements of the characteristics of the PWO scintillation elements, produced from the specified raw material, were also supplemented by the measurements taken in the particle accelerator at CERN in 1997–2001.

The following definitions were used in the preparation of the specifications:

The raw material is a mechanical mixture of the oxides of lead and tungsten for growing crystals by the Czochralski method in a controlled gas medium;

The quality is a number of physical and chemical properties of the charge essential for the production of crystals with the required parameters;

Concentration is the mass fraction (in %) of the content of any ion in the charge;

The impurity ion is the ion whose concentration in the raw material is not lower than 50 ppb.

The values taken into account in the development of specifications for the charge included the properties of the impurity centres in their location in the crystal, structural special features of the lead tungstate crystal, the classification of point structural defects and special features of doping essential for compensating the effect of these defects. The development of the specifications for the raw material of lead tungstate

Table 3.1. Distribution coefficient of the most important impurities

Chemical element	Effective distribution coefficient	Effect on important parameters of PWO scintillation elements
Na	1.2	Reduces radiation hardness
Si	0.4	Reduces radiation hardness
K	1.2	Reduces radiation hardness
Ca	1	Impairs scintillation kinetics
Cr	0.6	Impairs scintillation kinetics
Fe	0.8	Reduces radiation hardness
Cu	0.7	Impairs scintillation kinetics
Zn	0.1	Impairs scintillation kinetics
Y	1	Increases radiation hardness
Nb	0.68	Increases radiation hardness
Mo	1	Impairs scintillation kinetics
Cd	1.3	Impairs scintillation kinetics
Sn	1.4	Impairs scintillation kinetics
Ba	1.7	Reduces radiation hardness
La	1.1	Increases plasticity of crystals

[71] was preceded by the analysis of the effect of various impurities on the scintillation properties of the PWO crystals. The impurity ions appear in the raw material owing to the fact that they are present in the initial oxides or penetrate into the final product in purification and mixing, melting and synthesis, in storage or directly in the crystallisation process. The requirements on the concentration of the individual impurity ions and reasons for restricting the concentration of these impurities are shown in Table 3.2 [72].

In addition, the following factors were taken into account:

- the experiments with the variation of the stoichiometric composition of the raw material [73];
- experiments with preliminary synthesis of the raw material;
- special doping for increasing optical transmission and radiation hardness of the crystals and their light yield;
- the possibility of a flexible technological inspection and correction of the quality of the raw and secondary material in all production stages;
- ensuring the maximum safety in operation with the lead compound and fulfilment of the ecological standards;

Table 3.2. Requirements on the concentration of individual impurity ions and reasons for restricting their concentration

Element (s)	Effect on the properties of PWO crystals. Requirements on concentration
H	Hydrogen is present in the raw material because of the moisture content of the raw material. The negative effect of hydrogen on the properties of crystals has not been detected.
Li, Be, B, F	The elements of this group are found in the raw material because of the presence of these elements in the main lead oxide. The concentration of Li, Be, B, F in the raw material should not exceed 0.05 ppm and the total amount should not exceed 0.2 ppm
Na	In implantation in the lattice, Na substitutes the Pb ion and results in the localisation of the hole centres in the crystal. The negative effect of sodium on the radiation hardness of the crystal has been detected. The Na concentration of the raw material should not exceed 0.15 ppm. Prior to preparation of the raw material, the oxide should be subjected to purification.
Mg	In implantation in the lattice, magnesium substitutes the Pb ion and supports the formation of slowly emitting centres. The experimental results show the negative effect of this element on the luminescence kinetics of the crystal. The Mg concentration of the raw material should not be greater than 0.05 ppm. Prior to preparation of the raw material, the oxide should be purified.
Al	Aluminium is implanted in the interstitial positions in the structure of lead tungstate. The negative effect of this element on the properties of the crystal at the concentration smaller than 0.2 ppm was not reported. The aluminium concentration of the raw material should not exceed 0.2 ppm. Prior to preparation of the raw material, the oxides should be purified.
Si	Silicon is implanted in the lattice, substituting the W ion, and leads to the localisation of hole centres in the crystal. The negative effect of this element on the radiation hardness of the crystal and the resistance to mechanical effects has been reported. The silicon concentration of the raw material should not exceed 0.1 ppm. Special purification of the oxides prior to preparation of the raw material is essential.
K	Potassium is implanted into the lattice, substituting the Pb ion, and leads to the localisation of hole centres in the crystal. This element has a negative effect on the radiation hardness of the crystal and mechanical effects. The K concentration in the raw material should not be greater than 1.2 ppm. Special purification of the oxides prior to preparation of the raw material is essential.
Ca	Calcium is implanted in the lattice, substituting the Pb ions, and results in the formation of slowly emitting centres. Ca is always present in the lead oxide. The negative effect of this element on the luminescence kinetics of the crystal has been reported. The Ca concentration of the raw material should not exceed 3 ppm. Special purification of the oxides prior to preparation of the raw material is important.

Table 3.2. continued

Element (s)	Effect on the properties of PWO crystals. Requirements on concentration
Sc	Initially, scandium is not found in the raw material. No negative effect of this element on the crystal properties has been reported. The natural scandium concentration of the raw material does not exceed 0.05 ppm. No special purification is required prior to preparation of the raw material.
3*d*-elements	Because of optical transitions between the energy levels of the 3*d* shell, all the ions of the group from Ti to Ni has a negative effect on the kinetic properties of the PWO crystals. They substitute the W ion in the crystal lattice. The general requirement on the concentration is < 0.1 ppm, respectively. This limit does not include the requirements on the concentration of iron ions in the raw material. The negative effect of the presence of Fe on the radiation hardness of the crystals has been reported. The required Fe concentration in the raw material should not be greater than 0.07 ppm. Special purification of the oxides prior to preparation of the raw material must be carried out.
Cu, Zn	These ions are implanted in the crystal lattice, substituting the Pb ion, and leading to the formation of slowly emitting centres. The negative effect of both these elements on the luminescence kinetics of the crystals has been reported. The Cu and Zn concentration in the raw material should not exceed 0.08 and 0.2 ppm, respectively. Special purification of oxides prior to raw material preparation is essential.
Ga, Ge, Se	No systematic presence of these elements in the raw material has been detected. Similarly, these elements have no negative effect on the scintillation properties of the crystal, but the implantation of these elements in the crystal increases the brittleness of the latter. The concentration in the raw material should not exceed 0.05 ppm each.
As	Natural presence of As in the oxides has not been reported.
Y, Nb, Mo	The role of the ions of the group is described in section 2.2.
Rb, Sr, Zr, Te, Ru, Rh, Pd, Ag, In, Te, I	The presence of these elements in the raw material is not systematic. No negative effect of these elements of the properties of the crystals has been reported. The individual concentration of these elements in the purified raw material does not exceed 0.05 ppm. No special purification of these oxides to remove these elements is required prior to preparation of the raw material.
Cd, Sn	These ions, implanted in the crystal lattice, substitute the Pb ions and lead to the formation of slowly emitting centres. The individual negative effect of these elements on the scintillation kinetics of the crystals has been reported. The concentration of Cd and Sn in the raw material should not exceed 0.1 and 0.15 ppm, respectively. Special purification of the oxides prior to raw material preparation is required.
Sb	Natural presence of Sb in the oxides has not been detected. Nevertheless, it is obvious that the effect of this ion on the radiation hardness of the crystal is positive. However, it is not recommended to dope the crystal with Sb ions because of the toxicity of this element.

Table 3.2. continued

Element (s)	Effect on the properties of PWO crystals. Requirements on concentration
La, Lu, Ba	The role of these elements have been described in section 2.2. Because of the optical transitions between the energy levels of the 4f shell, all the ions of this group from Ce to Yb has a negative effect on the kinetic properties of the scintillations. They substitute the lead ion in the lattice. However, rare earth elements are naturally not present in the tungsten and lead oxides. The general requirement on the individual concentration of these elements in the raw material is 0.05 ppm. This limit does not include the requirements on the Ba concentration of the raw material. The required concentration of this element is 1 ppm.
Mo	The effect of molybdenum ions has been described in section 2.2. As a result of experimental growth trials with the crystals with different molybdenum concentration has been established that the Mo concentration in the raw material for the production of PWO crystals should not exceed 1 ppm. Special purification of the tungsten oxide prior to preparation the raw material is essential.
Ions with atomic number greater than 84, noble gases	Negative effect has not been detected. These elements are present in small concentrations in the oxides.

- dependence of the mechanical properties of the crystals on the composition of the raw material;
- the need for ensuring the maximum yield of suitable components in the mass production conditions;
- the possibility of efficient control of the quality of the raw material by acceptable chemical and physical methods;
- adherence to the European standards for the storage and transport of chemical reagents;
- need for ensuring long-term stability of the quality of the raw material and maximum protection against changes in suppliers of the raw oxides Pb and W;
- the minimum amount of the impurities with optical transitions;
- the minimum amount of the impurities generating cation and anion vacancies in the crystal.

The results of a large number of experiments show that:

- to ensure the required homogeneity of the charge, the initial oxides should be fragmented into the granule size of 10–30 μm;

Table 3.3. Main standards for lead tungsten charges

Parameter	Standard
Molar ratio Pb:W	1:1 +/–0.01
Mass fraction of Al, %, no more than	0.0001
Mass fraction of Fe, %, no more than	0.0002
Mass fraction of Si, %, no more than	0.001
Mass fraction of Mg, %, no more than	0.0001
Mass fraction of Mn, %, no more than	0.00002
Mass fraction of Cu, %, no more than	0.0002
Mass fraction of Mo, %, no more than	0.0005
Mass fraction of Cr, %, no more than	0.00005

- the tungsten oxide which brings the largest amount of the uncontrolled impurities into the composition of the charge should be chemically purified;
- the charge should be dried at a temperature of 100°C prior to packing into polyethylene bags;
- the moisture concentration in the charge prior to application should not exceed 0.1% in relation to the total weight;
- preliminary synthesis of the charge is not required;
- preliminary tableting or compacting of the charge is not required.

The main specifications norms for lead tungsten charges [72] are presented in Table 3.3.

The mass production of the raw material of lead tungstate was organised in Russia by the company Neokhim (Moscow) with the maximum productivity of 6 t/month.

3.2. Growth of PWO crystals

The main requirements on the mass production of lead tungsten scintillators are, on the one side, the need to retain the properties of the crystals within the framework of the specifications, for example, for CMS experiments, and on the other hand, the need to ensure the minimum production costs. It should be mentioned that in the mass growth of PWO crystals by the Czochralski method, these requirements are contradicting to a large extent.

The mass growth of PWO crystals was organised using the equipment available in Russia for crystal growth of the type Lazurit, Kristall-2 and Kristall-3M (Fig. 3.1). This is the equivalent of the same type fully automated for growing high-temperature oxide crystals by the

Fig. 3.1. Lazurit equipment for growing single crystals by the Czochralski method.

Fig. 3.2. General view of the chamber of growth equipment with the thermal section and a grown PWO crystal.

Czochralski method from the melt. In the growth process, the diameter of the growing crystals is controlled by continuous weighing and the heating and cooling algorithmes are also programmed, together with the speed of withdrawal of the crystal from the melt and the rotational speed of the bar. The chamber of growth equipment is designed and fitted in such a manner that solidification can be carried out in vacuum, in air or in inert gas. The dimensions of the thermal section enable PWO crystals weighing 2.7–2.9 kg to be drawn from a crucible with the melt with a mass of 11–12 kg (Fig. 3.2).

To ensure economic efficiency of the conversion of the raw material into intercrystalline mass (up to 80–85%), the number of consecutive solidification cycles from the same crucible by adding the raw material every time after the growth process was increased up to 20. However, the process of defect formation became more intensive with the increase of the number of solidification cycles in the crystals. The increase of the number of defects, both of intrinsic structural defects and impurity defects, impairs the parameters of the crystals, especially leads to the formation of slow components in scintillations and reduces the radiation hardness of the crystals. It has been established that the crystals with the parameters satisfying the specifications conditions of the CMS experiments can be produced by the process without 15 crystallisation cycles using the specific raw material and maintaining the stoichiometric composition of the melt during these crystallisation stages.

The lead tungstate crystals were grown in platinum or iridium crucibles in the atmosphere with the composition similar to that of air or in an inert gas medium [74, 75]. The raw material (charge), consisting of a homogeneous mechanical mixture of the oxides of lead and tungsten with the purity not lower than 99.999 %, is melted to increase the density and carry out preliminary synthesis. The raw material is recrystallised for the additional purification to remove uncontrolled impurities and for averaging the chemical and stoichiometric composition. In other words, the raw material for producing the boules from which the scintillation elements are produced, is represented by the already grown crystals. In subsequent crystallisation stages, the composition of the melt is constantly corrected and doping is carried out in accordance with a specific algorithm, taking into account the number of the solidification cycles and the results of measurement of the parameters of the elements produced from the previously made crystals. The doping level increases with the increase of the number of the solidification cycles. PWO crystals are doped with lanthanum and yttrium to the content of several tens of ppm and are grown in reproducible conditions. However, to correct the deviation in the parameters of the growing crystals, the growth conditions (for example, drawing rate, speed of rotation and the design of the thermal section) can be varied in a relatively wide range.

In addition to intensification of the processes of formation of the point defects with the increase of the sequence number of the solidification cycles in the crystals, the crystals are characterised by an increase of the amount of gas inclusions which greatly reduce the optical transmission and light collection in scintillation elements. To remove this undesirable phenomenon, the growing rate of the crystal from the melt decreases from 8–9 mm/h for the initial crystallisation cycles, to 5 and 6 mm/h for 13–15th crystallisation cycle.

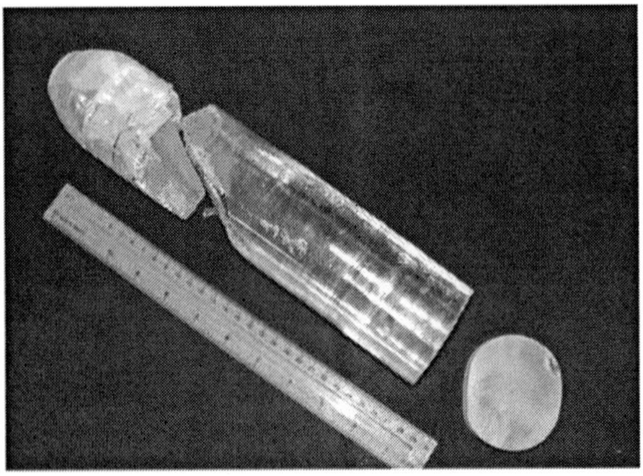

Fig. 3.3. Crystallisation boule which cracked during cutting. It is used as secondary material or material for producing seed crystals.

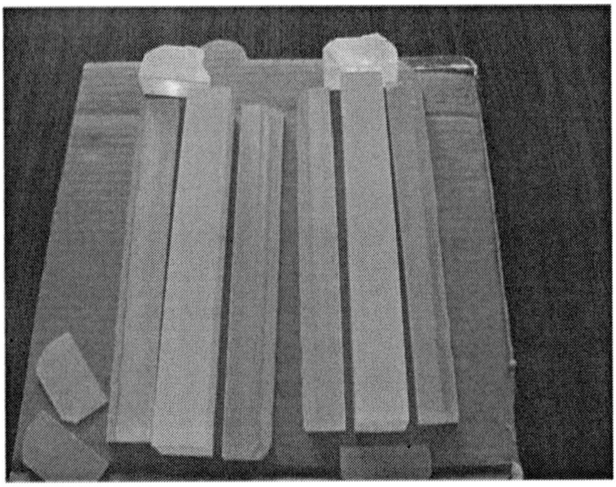

Fig. 3.4. The scintillation PWO elements cut from the crystals. The end and side parts are used as secondary material for growing new boules.

The mass production technology incorporates secondary application of the material – addition, to the raw material, of the rejected crystals (Fig. 3.3) and of the material after cutting the boules (Fig. 3.4). This was also taken into account in correcting the melt both with respect to the content of the doping impurities and the content of uncontrolled impurities and, on the whole, increases the utilisation factor of the raw material by at least 25%.

The boules are 290–310 mm long (with an accrescent cone) and their cross-section is elliptical, with the ellipse written into the circle with a diameter of 36–40 mm. Figure 3.5 shows tmass grown PWO

Fig. 3.5. PWO crystals in the machining section, grown over a period of 3 days and subjected to high temperature annealing.

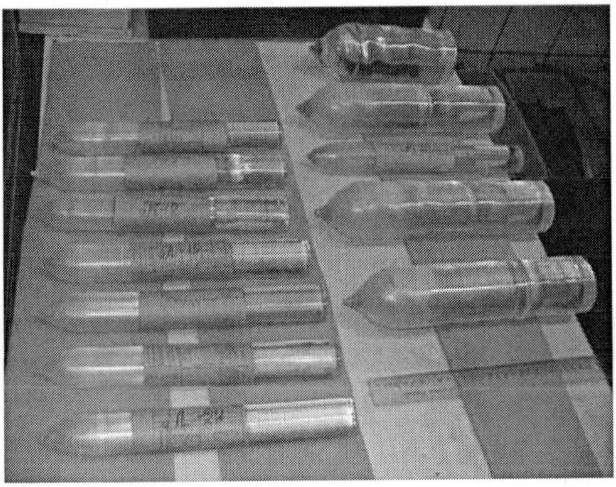

Fig. 3.6. Standard lead tungstate crystals (left) and experimental crystals with the diameter of 60 and 80 mm.

crystals. The crystals with a larger diameter, up to 80 mm, can also be grown (Fig. 3.6).

The most important growth conditions of the PWO crystals are:

- application of the stoichiometric mixture of WO_3 and Pb_3O_4;
- growth in the gas medium depleted in oxygen;
- the orientation of the seed along the crystallographic axis *a*;
- doping with La and Y ions to 50 ppm.

The stoichiometric mixture of WO_3 and Pb_3O_4 is used owing to the fact that, initially, in accordance with the industrial specifications, Pb_3O_4 was pure as regards microimpurities in comparison with PbO.

The crystals, grown in nitrogen or argon (the atmosphere is selected because of the higher cost of the gases) with the oxygen content of 10^{-3}–1 vol.% have the light yield 1.5 times higher than the crystals, produced in air. The growth direction along the *a* axis produces crystals with lower stresses, and the cleavage plane is situated along the crystal in this case. Doping the crystals with the La and Y ions helps to fulfil the requirement for the radiation hardness on the basis of the criterion of induced absorption not lower than 1.5 m^{-1} (see section 1.3.4) for 96% of the grown crystals. The yield of efficient scintillation PWO elements in mass production is more than 85%. The process of crystal growth is described in general in the patents of the Russian Federation No. 2132417 and 2164562 [74, 75].

3.2.1. Method of compacting and homogenisation of the raw material

The raw material for the growth of PWO crystals in the mass production conditions is a mechanical mixture of tungsten and lead oxides. The bulk density of the powder is ~5.0 g/cm^3. At this density, the crucible used for growing single crystals should be filled in 3–4 stages so that the growth cycle is extended by 20% or more. In addition, further thermal cycles (heating–cooling) greatly reduce the service life of crucibles. In the programme for ensuring quality at the Bogoroditsk Techno-Chemical Plant (BTCP), experiments were carried out with testing and introduction into service of an original method of melting the raw material (Fig. 3.7) which increases the density of the raw material up to 8 g/cm^3. In the method, the raw material is placed in a platinum crucible heated to the melting point and the melt is then poured into a platinum mould. Solidification of the melt results in the formation of tablets (Fig. 3.8) with the density close to the density

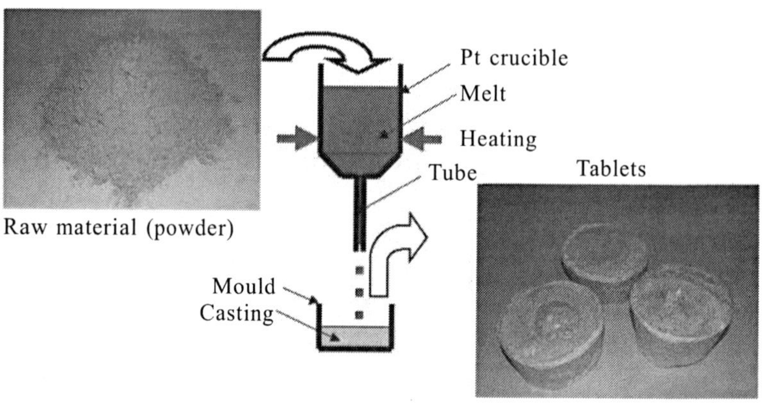

Fig. 3.7. The method of melting the raw material.

Fig. 3.8. Magnified photograph of a lead tungsten tablet.

Fig. 3.9. Equipment for melting the raw material.

of the PWO single crystal and with the diameter 5–6 mm smaller in comparison with the internal diameter of the crucible used for growth. The raw material is melted using Tesla 633 modernised equipment (Fig. 3.9). Equipment uses resistance heating and a crucible with the volume of approximately 5 litres. The productivity of equipment is 100 kg/day with continuous operation. The introduction of the technology of melting of the raw material released growth equipment from non-productive operations of additional melting of the raw material, i.e., the crucible is filled with the raw material in a single cycle immediately prior to growing the crystal.

3.2.2. Selection of crucible material and optimisation of crucible design

The efficient selection of crucibles for growing PWO crystals is a very important technical and economic task. Experience with service of various crucibles shows that the cost of the lease, processing and burning of crucibles of made of expensive metals is 20% of the production costs of PWO scintillation elements. The main initial requirements in the selection of the crucible material are:

- the cost and availability of the metal;
- the resistance of the metal to the effect of the lead tungsten melt;
- heat resistance of crucibles or resistance to deformation in thermal cycles;
- service life and efficiency of repairs.

Crucibles made of two metals (platinum and iridium) have been tested in the PWO growth technology. Comparative tests and experience with long-term service show that both crucible types are suitable for growing PWO crystals [74]. However, the service of iridium platinum crucibles has its own special features. In particular, iridium crucibles can be used in an inert gas or nitrogen. When using iridium crucibles in air, the metal losses are higher because of sublimation at the melting point of the lead tungstate. Experiments were carried out using iridium crucibles of two types: welded and plastic electroplated crucibles. The overall size of the crucible (diameter × height) was 130×140 mm, wall thickness 1 mm. Regardless of the crucible production method, the service characteristics of the crucibles were excellent. The mean service life for 27 iridium crucibles was 10512 h. Consequently, the crucibles can be used for growing PWO crystals for almost two years in the mass production conditions. For comparison, the mean service life of welded platinum crucibles according to the data for the years 2000–2005 was 8130 hours. The data were obtained from the results of tests on more than 300 crucibles. The data for the platinum crucibles do not include the data for failed crucibles, i.e., crucibles with hidden production defects or crucibles which failed in unsuccessful crystal growth processes.

In addition to evaluating the service characteristics, work was carried out to compare the parameters of scintillation elements produced from the crystals grown in Pt and Ir crucibles. The light yield, scintillation kinetics, transmission spectra and radiation hardness were compared. In the years 2000–2003, the parameters of more than 1500 crystals were measured. The radiation hardness is compared on 340 specimens. The results show that the main parameters of the crystals produced in the platinum and iridium crucibles are comparable. However, in the final analysis, the mass production of PWO crystals was organised using platinum crucibles because the infrastructure of secondary processing of these crucibles in Russia was more efficiently developed.

Another task for mass production of PWO crystals is the optimisation of the shape, dimension and design of the platinum crucibles. With a constant increase of the price of platinum, this task has become the main economic problem of mass production. Four technical solutions were gradually applied.

In the first years of mass production (1998–2000), standard welded platinum crucibles with the size of 120 × 130 × 1.5 mm (diameter × height × wall thickness) were used. In this period, the expansion of the scale of production was inhibited by restrictions on the amount and cost of platinum used in production.

In 2001, work started on the application of combined crucibles: platinum–ceramics [76]. Similar crucibles are used in glass production but the service conditions greatly differ from those in the crystal growth process. This applies to the multiple heating–cooling cycles. In the initial combined crucibles, the platinum wall was 0.6 mm thick, the thickness of the Al_2O_3 ceramic coating was ~3 mm. It may be seen that the thickness of the platinum wall of these crucibles is 2.5 times smaller than that of the standard crucibles. This also resulted in a significant saving economic benefit. However, the service life of the first two crucibles in 2001 was slightly more than 2500 hours which is three times shorter in comparison with the standard welded crucibles. The short service life was caused by the failure of the ceramic layer because of the excessively large thickness as a result of thermal cycling followed by disruption of the integrity of the platinum crucible (Fig. 3.10). Analysis of the initial results was used to formulate the requirements on the production of a pilot-plant batch. In a short period of time, the company Zavod Obrabotki Tsvetnych Metallov (Ekaterinburg, Russia) developed and improved two important technological processes for the production of combined crucibles: the technology of deposition of a protective coating on platinum, and a technology of producing seamless thin-wall platinum crucibles. The platinum crucible is the internal part of the combined crucible which is in contact with the melt and is produced in the seamless variant from platinum sheets 0.6 mm thick. Technology was developed for

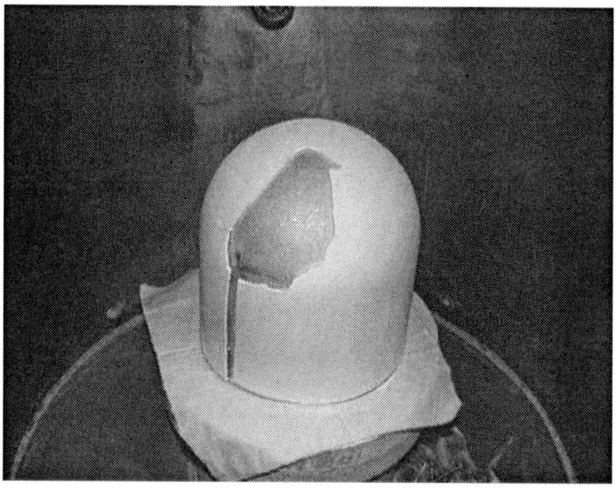

Fig. 3.10. Characteristic damage of the combined crucible ($Pt–Al_2O_3$) in service.

Fig. 3.11. Combined (platinum–ceramic) crucibles.

crucibles of all standard sizes, used in the production of lead tungstate crystals: 120×130; 130 × 140, 170 × 180 mm. In the years 2001–2003, BTCP company used 215 platinum–ceramic crucibles in production (Fig. 3.11). The mean service life was 4000 h. Individual crucibles reached a service life of 7000 h. There was no decrease or change in the parameters of the PWO scintillation elements. An obvious advantage of the combined crucibles is the decrease (by more than a factor of three) of the losses of platinum through sublimation. In addition, the wall thickness of the platinum crucible (0.6 mm) is optimum for using a transistor power source working at a frequency of 22 kHz.

It was found that even after complete destruction of the protective ceramic coating from the platinum crucible, the crucible can still operate efficiently for a certain period of time. This became the third stage of optimisation of the design of the platinum crucible – modification of the specifications for the conventional welded crucibles in part of the wall thickness. In 2003–2005, BTCP used in mass production platinum crucibles with a wall thickness of 700 μm. This 'bold' decision solved the problem of application of expensive platinum for a certain period of time.

New possibilities became available in 2004 when the company Supermetall (Zelenograd, Russia) developed a technology for production of a platinum-based laminated composite material (LCM). A thin wall sheet (500–700 μm) was in the form of a sandwich structure consisting of five alternating layers: platinum–composite–platinum–composite–platinum. The crucibles, produced from this material, showed excellent characteristics. The outer platinum layers prevented contamination of the melt with the container material and interaction of the crucible with the atmosphere during the growth of lead tungstate crystals. The internal layers of the composite materials provided the required rigidity to the thin crucibles which is very important in the thermal cycling conditions and in the case of large crucibles. Even at a wall thickness of 500–700 μm, the service life of these crucibles was longer than that of the standard crucibles with the wall thickness of 1.5 mm. The crucibles made of the LCM are still used in the mass production of PWO crystals.

As a result of tests of different types of crucible in the growth of PWO crystals, four types of crucibles have been introduced into production: iridium (welded and electroplated plastic), welded platinum, combined (platinum and ceramic), and crucibles made of LCM. They can all be used in the mass production of PWO crystals.

3.2.3. Modernisation of single crystal growth equipment

Experience with the mass production of PWO crystals at the Bogoroditsk Techno-Chemical Plant (BTCP) shows that the high energy requirement of production is a serious problem. In particular, this is caused by the application of outdated machine generators of VPCh-60-8000 type (60 kW, 8 kHz) for supplying electrical power to growth equipment. The generators were used in all growth equipment at the above plant. Critical shortcomings of the generators are: low efficiency, high losses in communication lines and non-optimum matching with the induction coil. This results in total losses of up to 50%. In addition, the generators are a source of strong noise and must be placed in separate, specially designed areas. Taking into account the yield of suitable crystals for the operation of crystal growth, the fraction of electrical energy in the production costs reached 40% (according to the data in 2003). For the planned operating life of up to 2000 hours, the losses due to the repair of generators also represent a significant part of the production costs. In the mean, the repair of a single generator costs US$700 per annum, taking into account only the expenditure on materials and component parts. Experience with the service of standard thyristor power sources for Kristall-3M equipment in the 80s was also negative. Because of the high hardness of water used for cooling the power sources, the operating life did not exceed 2000 hours and, consequently, the losses through service and repair were unjustifiably high.

With the appearance of IGBT power transistors on the market, the concept of construction of efficient power sources for growth machines became very important. In the course of mass production of lead tungstate crystals for ECAL CMS at CERN and the introduction of an energy-saving program, the main task was the substitution of machine generators by transistor generators (Fig. 3.12). Various types of power sources for growth equipment, including transistor power sources, are available on the world market. However, they are all expensive and require additional modification for matching with Russian growth equipment.

The contract for the development of the transistor power source was awarded to two companies: Antrel (Moscow, Russia) and

Fig. 3.12. Machine generator (left) and a prototype of a transistor power source (right) for Lazurit-type growth equipment.

ElektroSpetsPribor (Tula, Russia). The main technical requirements for the transistor power source (TPS) are shown in Table 3.4.

Comparative tests of two prototypes of the TPS produced by the two companies were carried out in 2003. The sources showed excellent service characteristics, exceeding the technical requirements. A saving of 50% of electric energy in comparison with machine generators was achieved. In January 2004, it was decided to re-equip the park of growth machines with transistor power sources. Preference was given to the transistor power source developed by ElektroSpetsPribor characterised by higher reliability and a shorter cost recovery time (1.5 years). Approximately 140 growth systems were re-equipped over a period of two years.

3.3. Annealing and machining of PWO crystals

After growth, the crystals are annealed in air in a Lantan low-gradient automated industrial furnace (Fig. 3.13) to reduce the mechanical stresses prior to treatment.

The volume of the working space of the services enables the group method of heat treatment to be used in which several crystals are annealed in a single process. The crystals are placed into alundum (Al_2O_3) and boxes onto a charge of fragmented PWO crystals to prevent mechanical contact and chemical interaction between the crystal and the base plate. At least 12 crystals can be annealed simultaneously at a

Table 3.4. Main technical requirements for the transistor power source.

No.	Parameter	Value
1	Maximum (nominal) output power, no less than, kW	20
2	Working frequency, kHz	20...24
3	Maximum output effective voltage, V	1000
4	Maximum output effective current, A, no less than	25
5	Power supply, V/Hz	3 phase × 380/50
6	Required power, no more than, kW	21
7	Voltage regulation range, V	0...10
8	Power regulation range, %	5...100
9	Permissible deviation of output power of the generator from the given power with Up variatioin of +10%, no more than, %	± 0.1
10	Mean running time to failure, no less than, hours	2000
11	Mean service life, no less than, years	5

Fig. 3.13. High-temperature Lantan furnace for annealing crystals.

temperature of 80–85% of the melting point of the crystal. The duration of the process is 4–5 hours, depending on the crystal diameter.

The scintillation elements are cut from the annealed boules in accordance with the specifications of the CMS experiments. The crystals are cut using precision grinding machines R5030SP produced by the LGB company (Fig. 3.14). Diamond discs, 2 mm thick, with an external cutting-edge are used for cutting. The cutting accuracy is 10 μm and the surface finish after cutting requires only one operation of fine grinding prior to polishing in mass production. Significant improvement of the cutting process resulted from the installation of two discs on the spindle of a machine and, consequently, the possibility of making two parallel cuts in a single cycle (Fig. 3.4). The quality of cutting and the yield of suitable crystals are not affected.

All the faces of the PWO crystal and elements are polished by the group method on plane grinding machines (Fig. 3.15).

3.4. Comments regarding the cost of PWO crystals

The mass production of lead tungstate crystals for the CMS experiments at CERN should be viewed from the viewpoint of general economic considerations. Here, we can see a classical example of monopoly, its variety – monopsony.

At the beginning of the 90s, a strategic solution on the construction of a new accelerator of charged particles of the Collider type was made in the area of high-energy physics by the international non-commercial organisation CERN. The aim of this experiment (initial budget US$4

Fig. 3.14. R5030SP grinding equipment.

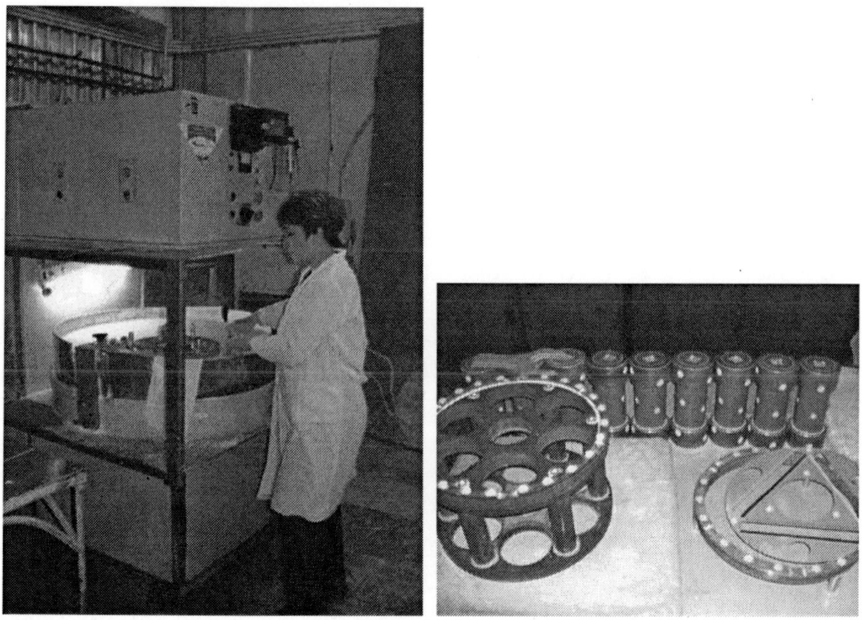

Fig. 3.15. Equipment (left) and jigging (right) for group polishing of PWO crystals.

billion) is the investigation of collisions of the proton and antiproton beams and, in particular, detection of the Higgs boson i.e., the decisive argument in order to confirm the Big Bang theory.

In the first stage, the CERN acted as the only customer for services and goods for the realisation of this unique project. This is a typical case of monopsony. Being in the unique position and also restricted by the budget considerations, the CERN naturally tried to minimise the cost of development of new materials, devices, equipment, etc. Nevertheless, a large number of companies took place in the project by submitting tenders. Consequently, a limited number of firms was accepted as suppliers, and only single firms were selected in the cases of some products.

A suitable example of the scenario is the production of crystals – PWO scintillators. In 1992, the CERN announced a tender for the development of a new crystal – scintillator, which was not available on the market at the time. The technical solution for this development greatly exceeded all the best specimens of crystals – scintillators available in the world at that time as regards the parameters and properties. Six companies from five countries (two of them were Russian) took part in the initial stage. In 1994, on the basis of the results of comparative tests, a final selection was made in favour of the lead tungstate crystal, where the technology of mass production was developed by experts of the BTCP company (Russia).

Utilising its unique position, the CERN greatly economised on the cost development because the competing companies in the effort to win the tender used their natural resources or resources of their governments because of the prestigious nature and potential of participating in the project.

Thus, in the resultant situation there was only one user and actually only one seller (the participation of Chinese companies did not exceed 8% of the number of produced crystals). It is interesting to examine the dynamics of variation of the prices of the product. As already mentioned, in the first stage, the user dictated the prices for the development and supply of experimental and small-series batches of the crystals. The price of the crystal was 2–2.5 US$/cm^3. In the second stage when the monopoly producer of the crystals was found, the prices became to a large extent the subject of 'haggling' between the user and the seller. This was clearly evident in the final stage of production when the supplier could secure by 'haggling' suitable conditions for themselves for completing the project. These conditions did not cover the considerable delays suffered by the company in the inital stages. The supplies were completed in 2008 and the price of the crystals was already US$5–5.8/cm^3. Thus, a relatively rare case of monopoly: monopoly user–monopoly producer, was realised in the mass production of the tungstate crystals for the CMS experiments at CERN.

After all, this is a unique case and is not at all characteristic of industrial production. However, in the case of crystals–PWO scintillators this was not the final stage. Positive experience with the construction of unique physical experiments at CERN became the subject of replication for other scientific centres. For example, physicists in Germany, USA and Japan decided to construct in their research centres new generation detectors based on lead tungstate crystals. A new situation formed in the market. Several new users of the PWO crystals arrived on the market. The PWO became some sort of standard in the area of high-energy physics for the next 10–15 years. The monopoly position of the Russian company was diminished and competitors again started to fight for the market.

In the new situation, it is not easy to predict the scenario of development of this competition. It can be assumed that the situation is being repeated to some extent. The reasons for this and advantages of this are evident, and the basis was formed by the previous 10 year experience with monopoly production. The BTCP company actually has the know-how as a result of being the only supplier of the raw material for the production of high-quality crystals. Utilising the inaction of competing companies, the BTCP has acquired considerable technological expertise and operates efficient production systems, without any need for large capital investment which is required by new 'entering' companies, i.e., there are barriers formed by the classic scheme of action of monopoly companies.

The conclusions are evident. Firstly, the monopoly, including 'pure' monopoly, is not the only economic category at the present time but it does exist and still has a strong effect on economic life. Secondly, the companies which have 'tasted' the market advantages of monopoly (and also monopsony) will naturally try to maintain and strengthen this position.

3.5. General characteristics of the mass production of PWO crystals

To complete this section concerned with the technology of mass production of lead tungstate crystals, it is useful to mention the most general factors:

- the method of production of crystals – the Czochralski method;
- the number of growth systems – 160;
- maximum productivity of the production line – 1200 elements (1.3 tonnes or 0.16 m^3) per month;
- the yield of suitable scintillation elements PWO – no less than 85%;
- the number of personnel 300 (administrative, engineering, basic and auxiliary);
- duration of production 10 years;

- the main technological stages;
- melting of the raw material;
- the primary recrystallisation of the raw material;
- growing PWO single crystals;
- annealing of the PWO single crystals;
- optical–mechanical treatment of the crystals;
- certification of scintillation elements.

4. A quality control system in mass production of lead tungstate crystals

4.1. Special features of certification of PWO scintillators for the CMS project

The production of lead tungstate scintillators for ECAL CMS is mass production system which has been operating at the maximum productivity for 10 years in order to produce more than 75 000 PWO crystals without storing the products. This means that a breakdown in production of crystals could have a critical negative effect on the entire course of the CMS project. Consequently, the production and delivery of crystals had to be accompanied by an efficient system of measures ensuring the efficient production of scintillation crystals with the required properties and a minimum number of complaints.

The data on the physical properties of the scintillation crystals of lead tungstate obtained in the investigations have been used to formulate the specifications for the parameters of crystals and determining the sequence of essential measures carried out prior to the production of crystals, during production, delivery of crystals and service.

All the essential measurements can be divided into two groups: certification and control measurements [77, 78].

Certification measurements consist of a set of measurements which enable the customer to receive components. This set includes specifications for the geometrical, optical and scintillation parameters (radiation spectrum, optical transmission at different wavelengths, measured along and across the crystal axis, light yield, the scintillation kinetics and nonuniformity of light yield).

Control measurements consist of a set of measurements which enable the producer to control the reproducibility of the properties of lead tungstate crystals throughout the entire production period. In fact, the control measurements should provide reliable information for understanding the evolution of production technology.

4.2. Methods and means of certification and control measurements of the scintillation parameters of PWO crystals

Optical transmission measurements

The typical transmission spectra of the PWO crystal consist of a plateau in the range 500–700 nm and a smoothly decreasing transmission curve in the range below 500 nm. This means that the transmission spectrum of the PWO does not have narrow absorption bands so that it is not necessary to take measurements of optical transmission with a high spectral resolution. The transmission spectrum can be measured at several selected wavelengths so that the measurement time can be greatly reduced. 11 wavelengths were selected for the investigations: 330, 340, 350, 360, 380, 405, 420, 450, 492, 620, 700 nm. This approach was realised in the ACCOS system (section 4.4).

Measurements of light yield, light yield nonuniformity and scintillation kinetics

The light yield (LY) and the light yield nonuniformity (LYN) can be determined by, for example, recording the position of the peak of total absorption in the amplitude spectrum of the ^{60}Co isotope sufficiently separated from the Compton signals and the noise of the photoelectric multiplier. This method is used for selection control measurements of the light yield.

Because of the relatively low light yield of the PWO (5–20 photoelectrons/MeV, i.e. 6–24 photoelectrons in the peak of total absorption of the gamma quanta with the mean energy of 1.25 MeV, source ^{60}Co), it is necessary to use a reflecting coating of the crystal and an immersion medium between the crystal and the photoelectronic multiplier. These requirements greatly reduce the flexibility and speed of the measurements and introduce an additional systematic error. The method is suitable for flow line investigations with a load of 2–3 crystals/h.

The ACCOS system uses a method of counting the photons which is combined with the measurement of the scintillation kinetics by the start–stop method. Excitation is carried out using a source of annihilation gamma quanta ^{22}Na.

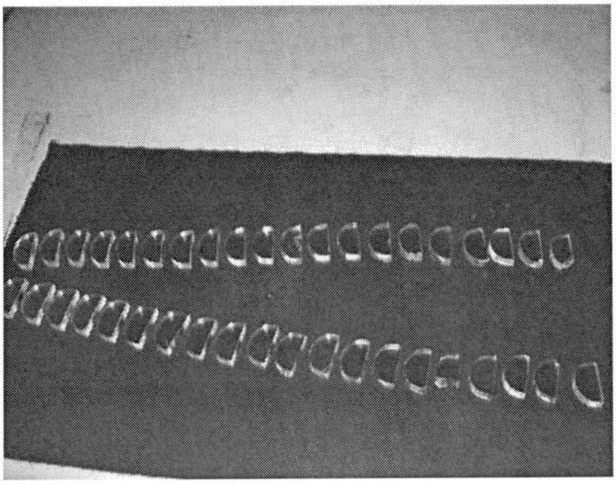

Fig. 4.1. Upper parts of the PWO crystals for induced absorption measurements.

Investigations of radiation hardness

It is not possible to measure the radiation hardness of all the supplied crystals because of the extremely high labour content and the large distance between the specific equipment and the production areas. The radiation hardness of the crystals is evaluated on the basis of the correlation between the value of the radiation hardness of the crystals and the angle of inclination of the edge of the transmission spectrum, measured along the axis in the wavelength range from 340 to 380 nm. At the same time, it is efficient to take additional measurements of the radiation hardness of part of scintillation crystals using the following procedure:

a. measurement of induced absorption in the upper sections of the crystals (Fig. 4.1) for the first (1–2) and last (14–15) solidification cycles in transition to new batches of the raw material. This method is used for selection control measurements.
b. selective verification of the full-size crystals using a radioactive source with measurements of the induced optical absorption or the change of the signal from a beam of high energy particles.

The methods of measurement of radiation hardness are described in detail in section 4.4.

4.2.1. Algorithm of selection of scintillation crystals for application in electromagnetic calorimetry in mass production

All the essential measurements, both certification and control measurements, are taken within the framework of the control flow diagram, shown in Fig. 4.2.

The diagram seems the participation of both the producer and user of the crystals. On the other hand, this interaction in the mass production process has made it possible to formalise the levels of accepting solutions, whose hierarchy shown in Fig. 4.3.

As indicated by these diagrams, in the first level, directly at the BTCP plant, laboratory means of measurements and the system of

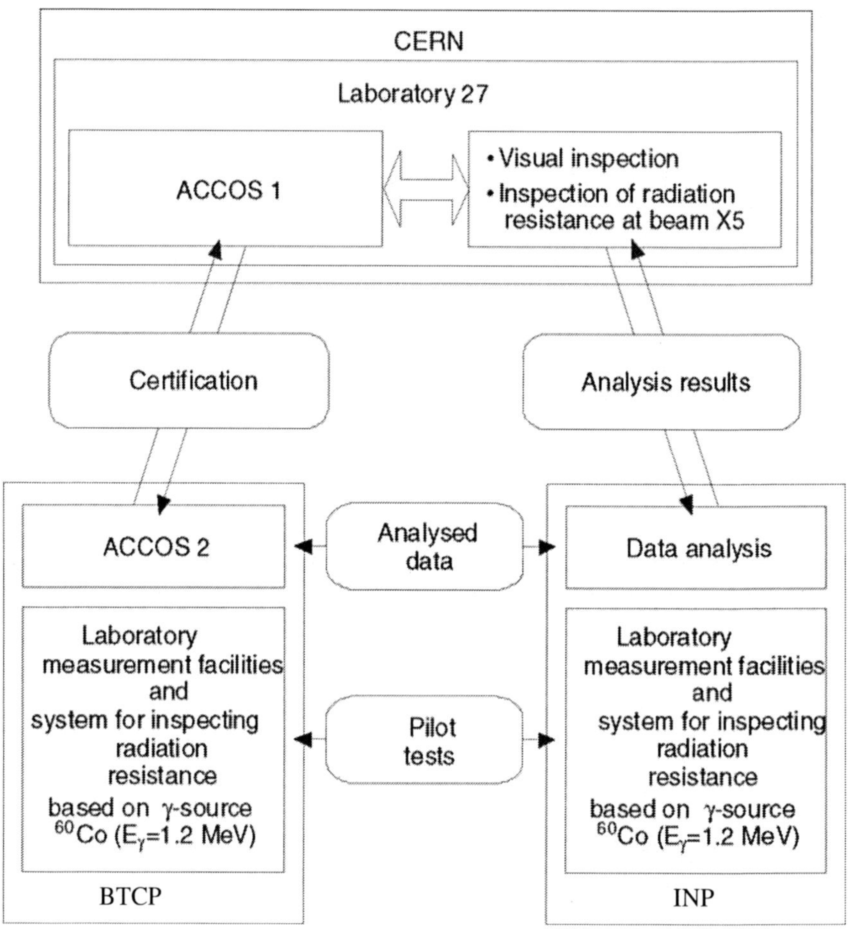

Fig. 4.2. Flow chart of control of the parameters of the PWO crystals.

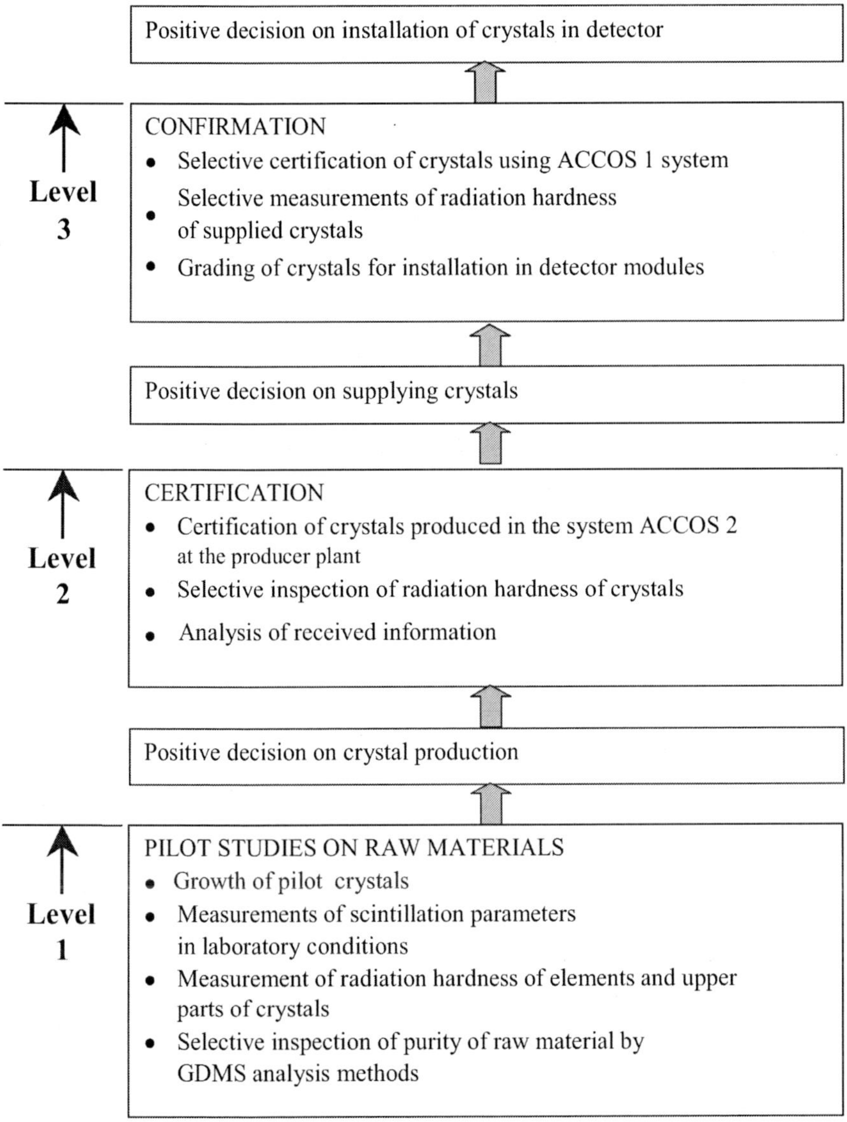

Fig. 4.3. Hierarchy of the levels of taking decisions in control of the parameters of the PWO crystals.

control of radiation hardness of crystals using the ^{60}Co source are used for investigating pilot batches of crystals and also for selective control of the raw material by the GDMS method. Subsequently, the Research Institute of Nuclear Problems (INP), being an independent expert organisation with essential analytical equipment and qualified experts, carries out analysis and selective verification of the data. If the results are positive, the given batch of the raw material is accepted for crystal production.

On the second level, the crystals of the produced batch are certified at the plant by the ACCOS-2 system, together with selective inspection of the radiation hardness of the crystals. The certification results are analysed at the INP and sent to the plant and, consequently, a decision is taken for every crystal as to whether it can be sent to CERN. The INP sends to CERN the results of analysis of the crystals which were accepted for dispatch to CERN.

On the third level, Laboratory 27 at CERN carries out registration and visual inspection of the supplied crystals. Subsequently, the ACCOS-1 system, which is identical with the ACCOS-2 system, is used for the certification of the crystals and also for selective testing of crystals in a particle beam in the X5 complex. Subsequently, the crystals are graded in accordance with their parameters. The results are analysed in Laborarory 27 and the INP. On the basis of the analysis results, CERN makes a decision regarding every crystal as to whether it is suitable for installing in the modules of the detector.

Thus, the proposed scheme of interaction of the parties and efficient exchange of the data should ensure successful production of crystals over a long period of time.

4.2.2. ACCOS – Automatic Crystal Control System

The crystal control system suitable for application in the mass production conditions should be automated to the maximum extent in order to reduce the effect of the subjective factor and have the capacity of at least 40 crystals in 24 hours or 1200 crystals in a month.

For mass production, The INP (Minsk, Belarus) together with the Particle Physics Laboratory at Annecy (LAPP, Annecy, France) and CERN (Geneva, Switzerland) developed the Automatic Crystal Control System (ACCOS). The general view of equipment is shown in Fig. 4.4.

Since the PWO crystals are relatively brittle and may contain mechanical damage, it is necessary to minimise the extent of handling the crystals. A suitable solution is to place crystals on special stands situated on a rotating platform. The crystals remain stationary in relation to the platform throughout the entire measurement cycle and the spectrometers move along the measured crystal. This solution is required development of appropriate compact spectrometers. The devices developed for this purpose also ensure accurate positioning of the crystals in relation to the measuring devices.

Fig. 4.4. Automatic Crystal Control System.

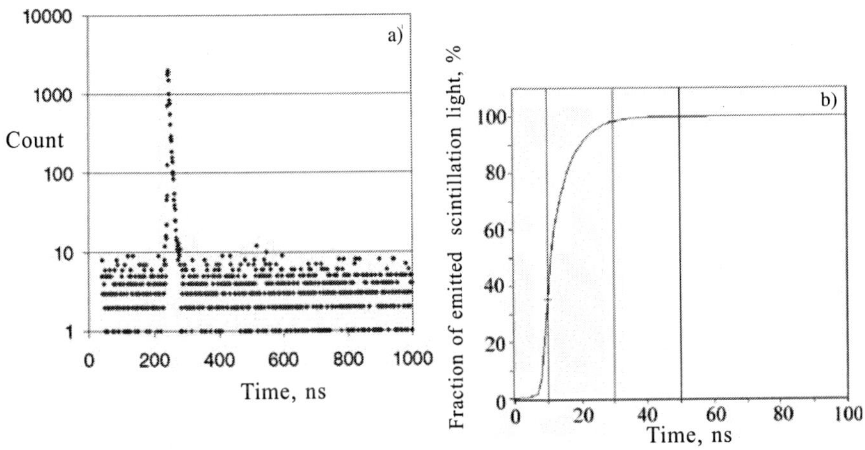

Fig. 4.5. Typical scintillation kinetics of PWO crystal (a) and its integral (b).

Figure 4.5a shows the typical scintillation kinetics of a PWO crystal, Fig. 4.5b is its integral, calculated using the analytical software of the ACCOS system.

Figure 4.6 a, b shows the typical spectra of optical transmission of the PWO crystal measured along (a) and across (b) the crystal axis. The graph shows the scatter of the values of the wavelength $\Delta\lambda$ of the transmission spectra, measured across the crystal axis, and the angle of inclination of the edge of the transmission spectrum in the wavelength range from 340 to 380 nm, measured along the axis.

In the stage of development of technology and production of the experimental batches it was found that there is a good agreement

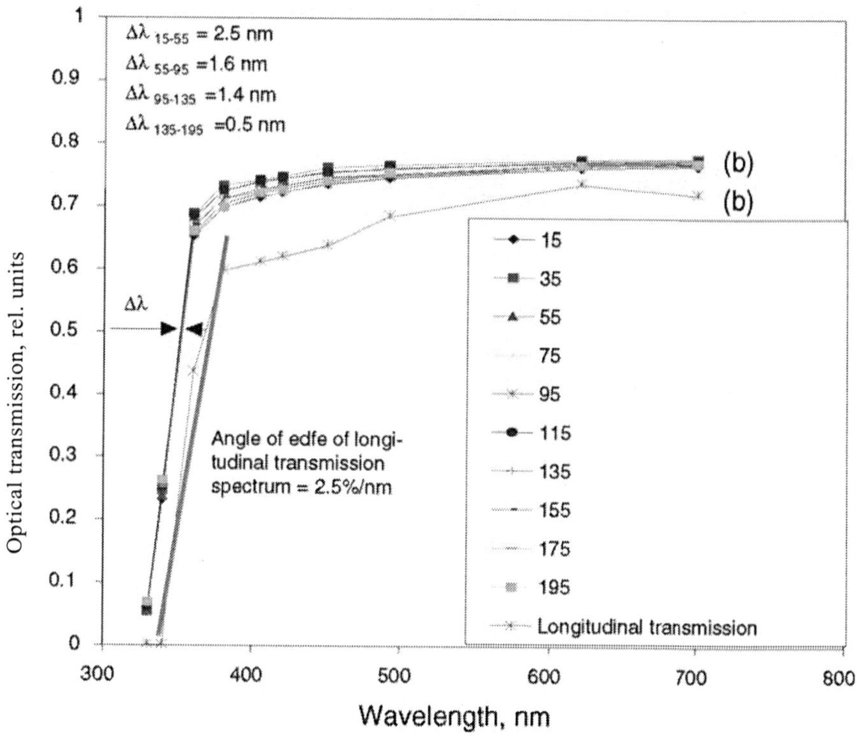

Fig. 4.6. Typical optical transmission spectra of a PWO crystal measured along (a) and across (b) the crystal axis.

between the results obtained using the ACCOS system and conventional laboratory investigation methods. Thus, it was concluded that the proposed approach to the certification of the PWO crystals and equipment used for this method in the form of the ACCOS system can be used for reliable certification of the scintillators for completing the ECAL CMS detectors [81, 82].

4.2.3. Methods and means of measuring the radiation strength of PWO crystals

Different radiation centres taking part in the investigations of the radiation strength of the scintillation elements on the basis of lead tungstate, the types of sources and the radiation doses are listed in section 2.1.

The system for controlling the radiation hardness of the crystals in the mass production conditions must satisfy the following conditions:

- must be suitable for testing full-size crystals (230 mm long) or specially prepared specimens;
- the variation of optical transmission should be measured at fixed wavelengths and over a short period of time;
- as a result of the dependence of the radiation-induced damage to optical transmission by radiation intensity, measurements should be taken in the actual conditions, i.e., at the radiation intensity of 15–100 rad/h, corresponding to the actual experimental conditions planned in the CMS project. This requirement indicate that it is necessary to develop a system suitable for long-term continuous operation (seven days) in the automatic regime;
- the system should enable testing of several crystals at the same time.

The RGB spectrometric system was developed for monitoring the radiation hardness and the restoration time constant of the PWO crystals [83]. This system has the form of an apparatus-software complex ensuring efficient automatic measurement of the decrease of the optical transmission of the PWO crystals under the effect of ionising radiation followed by automatic computation of radiation-induced absorption.

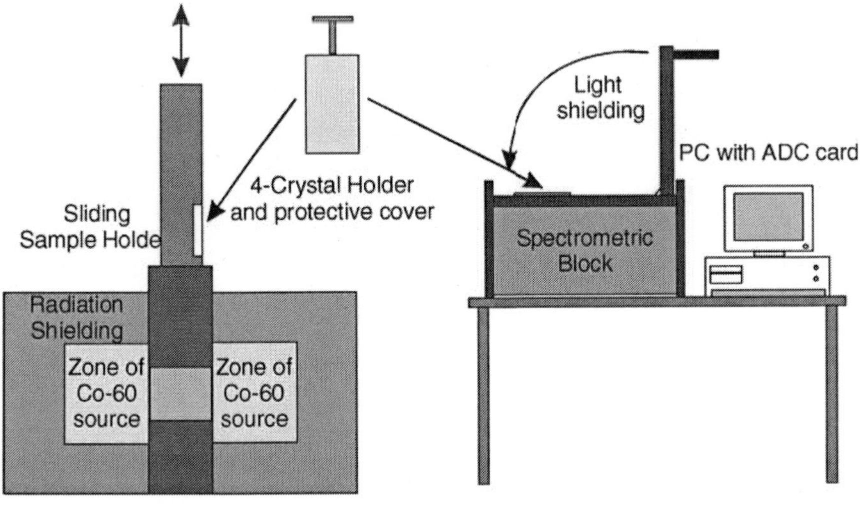

Fig. 4.7. Spectrometric system in the radiation centre.

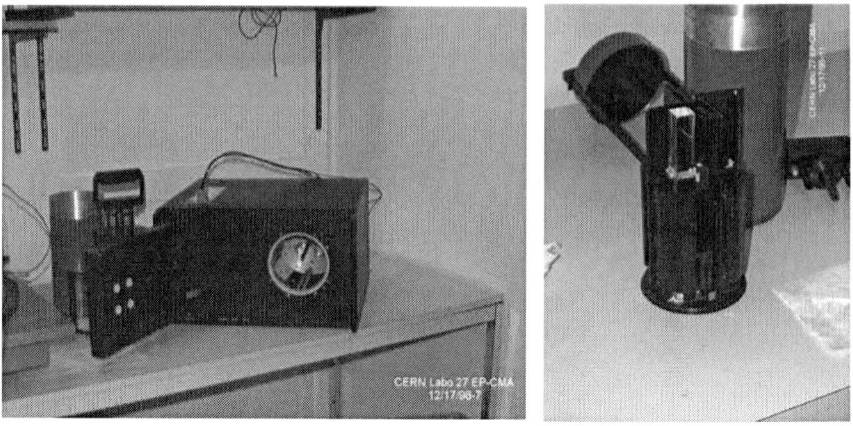

Fig. 4.8. Left – spectrometric unit (the lid of the cassette holder is opened). Right – a four-position cassette with a single crystal in it.

The system consists of a measuring unit with a cassette and a computer with special software. Four PWO crystals can be placed in the cassette.

The spectrometer measures the variation of the optical transmission along the axes of a PWO crystal after irradiation at three wavelengths: 630 nm (RED), 525 nm (GREEN) and 460 nm (BLUE). Because of the three colours, the spectrometric system is referred to as the RGB system.

The system is fitted with a ^{60}Co radioactive source with circular geometry and placed in a block with radiation protection. A cylindrical internal container with four PWO crystals, placed in a shaft, is irradiated in the transverse direction from all sides with a power of the order of 10 krad/h for 10 min. The photographs of equipment are shown in Fig.4.7 and 4.8. The device is fully automated. After activating, auto-calibration is started and this switches on the regulation of the number of photons in the channel and the intensification level in each channel. This procedure is always repeated twice – immediately after activation and after heating for 20 min. The measurement procedure includes the initial measurement of optical transmission, extraction of the internal container from the spectrometer and placing it in the shaft, irradiation, placing the container back into the spectrometer and measuring of damage and restoration of the optical transmission of the crystals. Since transmission can be measured with high accuracy only for several minutes, the time constant of parasitic restoration can also be measured. The long-term stability of the system is $\pm 0.1\%$.

The software developed for this purpose calculates directly the induced absorption in the crystal instead of the relative variation of the signal of the light diode (variation of the light yield).

To determine the correlations between the radiation data obtained in different radiation conditions using different methods and equipment, measurements were taken both at BTCP and other radiation centres. The measurements showed the satisfactory linearity of the correlation of the results obtained using the RGB system with the results of measurements obtained in other systems. Consequently, the RGB system was accepted in the CMS project as the main system for control by the producer of the radiation hardness of mass-produced PWO crystals [84, 85].

4.2.4. The method of controlling micro-impurities in crystals

The parameters of the raw material can be controlled by direct methods of chemical-analytical control or indirectly by direct measurement of the parameters of pilot crystals grown from every batch of the raw material. Prior to using every batch of the raw material in mass production, the following parameters must be determined with a high degree of reliability:

- the correspondence of the unity of the raw material to the requirements imposed on the groups of elements (elements of the iron group, rare-earth elements, molybdenum);
- correspondence of the raw material to the technical requirements for the grain size;
- correspondence of the raw material to the requirements for the presence of organic compounds and moisture.

In principle, all three parameters can be verified by combining the GDMS (glow discharge mass spectrometry) method, optical microscopy and chemical analysis. The second and third measurements can be taken independently by the producer, whereas the first measurement can be carried out reliably by a specialised analytical company. However, taking into account the distance to the reliable GDMS measurement unit, the expert analysis for the raw material can be delayed by 2–3 weeks which is unacceptable in the mass production conditions.

Therefore, a method was applied in which small high-density crystals are grown. In the mass production conditions, the procedure lasts 2–3 days using a single measurement system. Only 7–8 growth processes are required in the production of up to 1200 crystals per month in supplying raw material in batches of up to 400 kg.

The results of measurements of optical transmission, scintillation kinetics and the radiation hardness of the probe crystal are used for the reliable determination of the quality of the raw material.

Optical transmission provides information on the presence in the crystal of elements of the iron group, in particular, iron in the bivalent and trivalent states.

Scintillation kinetics provides information on the presence of molybdenum in the raw material.

Measurements of radiation-induced absorption are used to determine which centres in the crystal are damaged. Comparison of the shape of the spectrum and the amplitude of induced absorption at specific wavelengths (375, 420, 550, 620 and 720 nm) in the pilot and reference crystals enables reliable determination of the presence of heterovalent impurities, such as Eu, Yb, Fe, Cu.

4.3. Statistical analysis of the results of certification of crystals

4.3.1. Analysis of the distribution of the main crystal characteristics

In 1993–2093, more than 20 000 PWO crystals were produced at the BTCP for supply to CERN. Prior to supplying the crystals, measurements were taken of the scintillation properties of these crystals. Parameters such as light yield, scintillation kinetics and radiation hardness were analysed. Radiation hardness was estimated on the basis of the results of measurements of radiation-induced optical absorption at a wavelength of 460 nm. The light yield and scintillation kinetics were investigated for all crystals. Radiation-induced absorption was inspected for more than 7000 crystals. The data were analysed in detail. The dependence of the scintillation characteristics of the crystals on the number of the solidification cycle on the technological accuracy of every growth equipment was analysed.

The results of measurements of the crystals are presented in Fig. 4.9–4.18. The data show satisfactory technological accuracy of growth equipment and reproducibility of technology in the mass growth of lead tungstate crystals.

4.3.2. Analysis of the parameters of statistical distribution of the radiation hardness of PWO crystals

The parameters of the statistical distribution of the radiation hardness of PWO crystals in the radiation conditions of service in ECAL CMS were evaluated using the results of certification measurements [20],

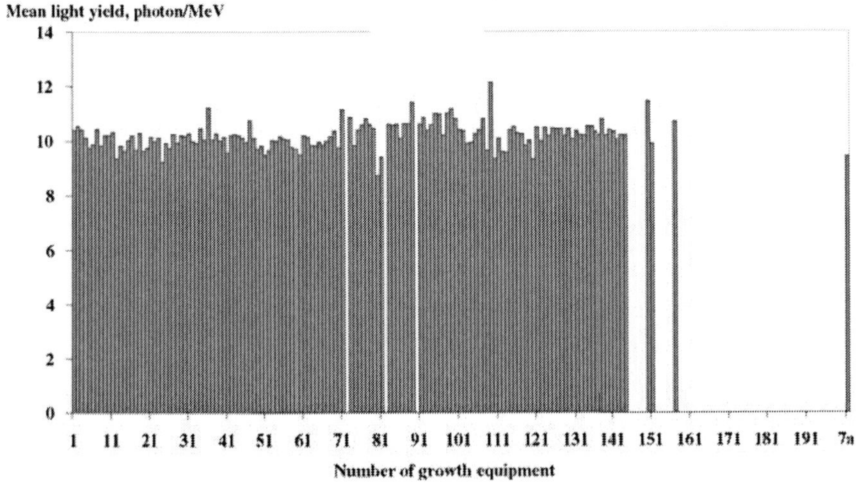

Fig. 4.9. Distribution of the mean light yield of crystals in relation to the number of growth equipment, photoelectron/MeV (15 346 crystals).

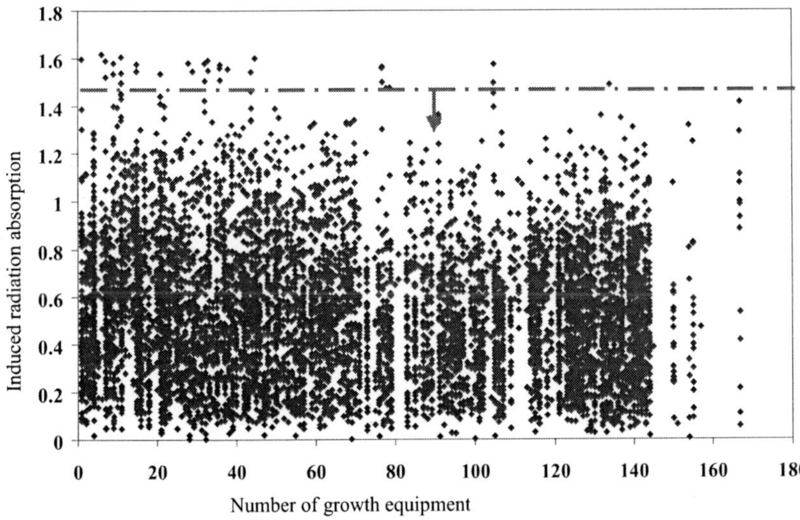

Fig. 4.10. Distribution of radiation-induced absorption of crystals in relation to the number of growth equipment (7633 crystals), the critical value $\mu < 1.5\ m^{-1}$.

taken at the BTCP in irradiation with a saturating dose (source of γ-radiation ^{60}Co, E_y = 1.2 MeV, dose power > 100 Gy/h, dose 40 Gy) and the results of experimental investigations of radiation hardness obtained at the GIF CERN for the working radiation conditions (source of γ-radiation Cs^{137}, E_y = 0.66 MeV, power dose 0.15 Gy/h, radiation dose 6Gy) [83, 84, 87].

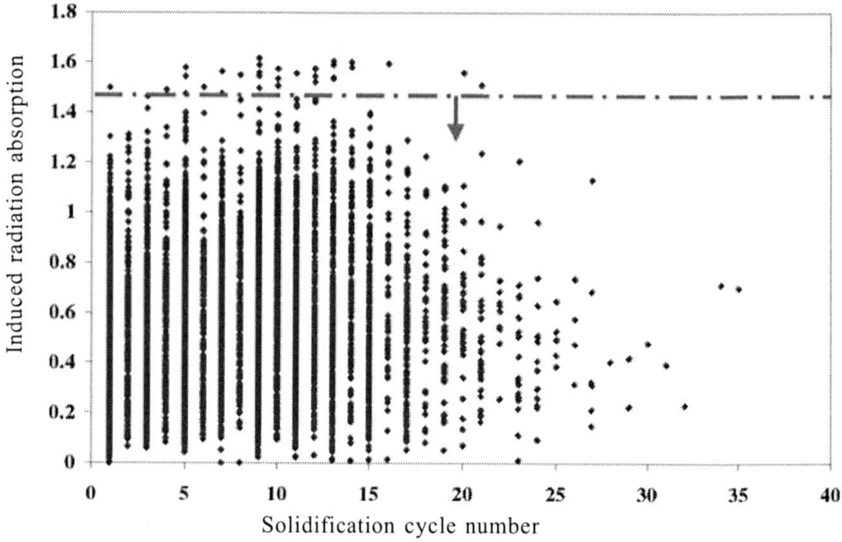

Fig. 4.11. Distribution of radiation-induced absorption of the crystals in relation to the crystallisation number (7633 crystals), critical value $\mu < 1.5\ m^{-1}$.

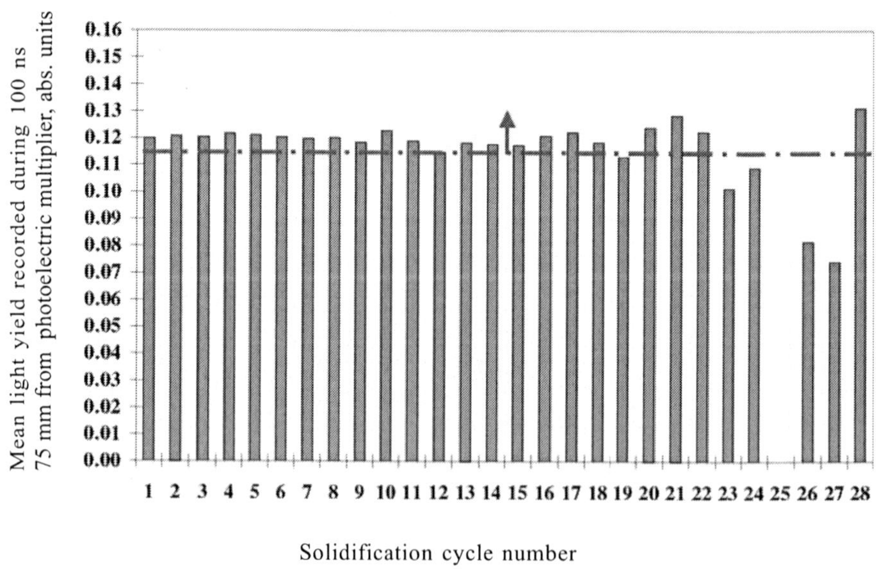

Fig. 4.12. Distribution of the mean scintillation light yield in the range 100 ns in relation to the solidification number (4112 crystals).

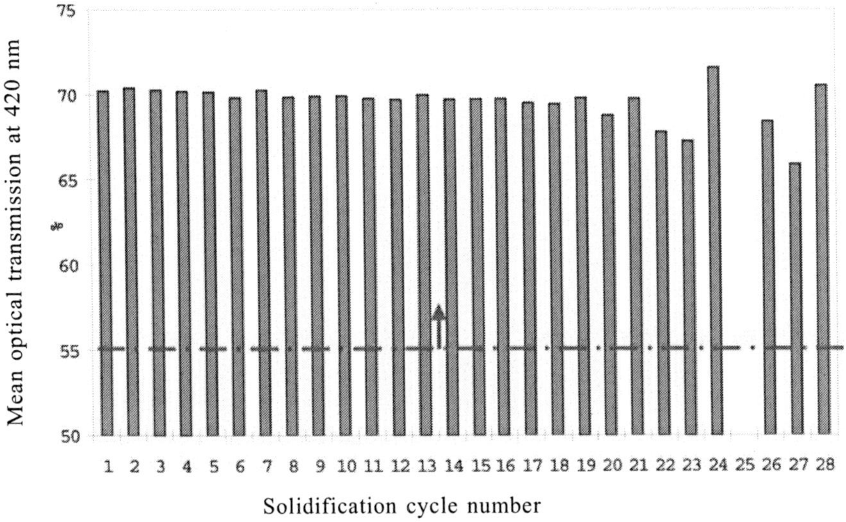

Fig. 4.13. Distribution of the light transmission of the crystals at 420 nm in relation to the solidification number (4112 crystals). Critical value $T = 55\%$.

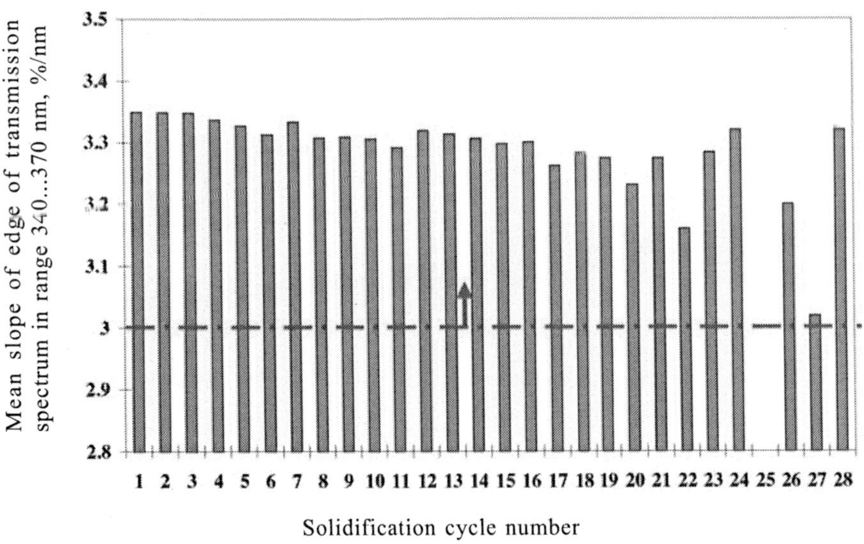

Fig. 4.14. Distribution of the angle of inclination of the edge of the spectrum (340... 370 nm) of longitudinal transmission (%/nm) in relation to the solidification number (4112 crystals). Critical value $S = 3$.

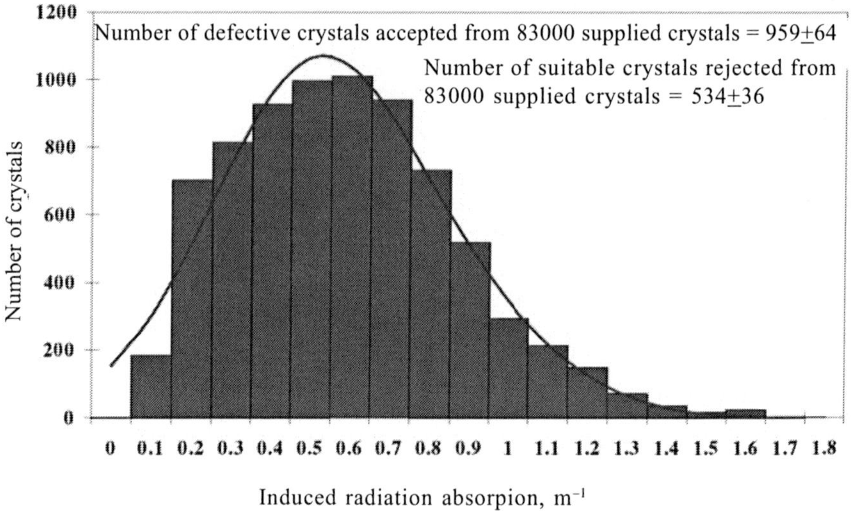

Fig. 4.15. Distribution of crystals on the basis of the value of radiation-induced absorption (7633 crystals). The critical value $\mu = 1.5\ m^{-1}$).

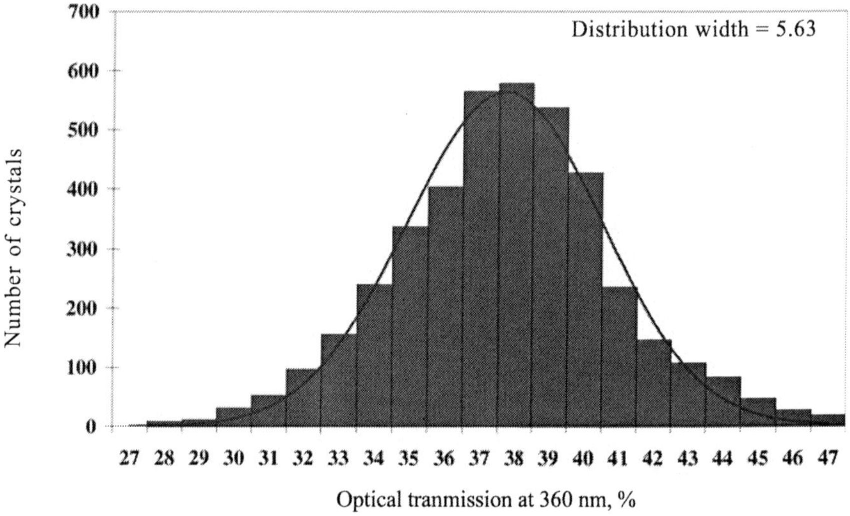

Fig. 4.16. Distribution of the crystals in relation to transverse optical transmission at 360 nm (4112 crystals).

The radiation hardness of the PWO crystals is determined by measuring the relative optical transmission of the crystals δT in radiation:

$$\delta T\ (\%) = (T_0 - T_1) \tag{4.1}$$

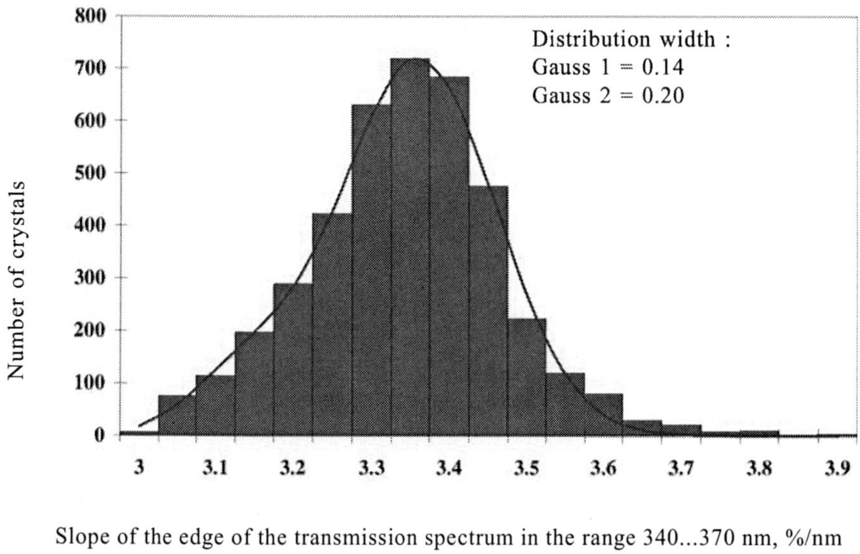

Fig. 4.17. Distribution of the angle of inclination of the edges of the spectra (340-s 70 nm) of longitudinal transmission (%/nm) (4112 crystals).

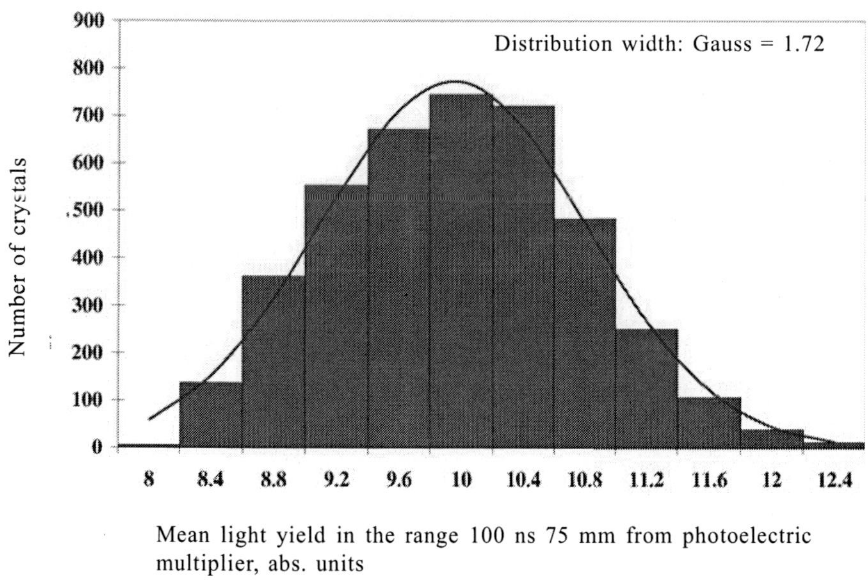

Fig. 4.18. Distribution of crystals on the basis of the mean scintillation light yield in the range 100 ns (4112 crystals).

where T_0 and T_1 in % are the values of optical transmission prior to and after irradiation in relation to the reference channel, regarded as 100%.

The technology of production of radiation hard PWO crystals was developed in 1998–1999. This was followed by investigations on two batches of crystals: the first batch consisted of 6594 crystals, produced in 1999–2001, and the second batch consisted of 8754 crystals, produced in 2002–2003. The summary data on the conditions and volumes of the investigations, the measured parameters and the relationships determined for obtaining estimates of the statistical parameters of the radiation hardness of the batches of PWO crystals are presented in Table 4.1*a* and *b* [83, 85, 86]. To evaluate the parameters of the statistical distribution of the photoelectronic term a_m of the electromagnetic calorimeter, the initial data sets included the results of measurement of the photoelectrical yield N_{phe} for the first and second batch, produced at the BTCP. The value of the photoelectric term for the PWO crystals a_m (%) at an energy of 1 GeV in relation to the measured value of N_{phe} (photoelectron/MeV) are determined from the following equation [19, 20]:

$$a_m = 6.5/\sqrt{N_{\text{phe}}} \qquad (4.2)$$

The mean value of the photoelectric term $<a_{mI}>$for the first batch of the crystals was 2.096%, and the mean quadratic deviation $\sigma(a_{mI})$ was 0.108%, and for the second batch $<a_{mII}> = 2.01\%$ and $\sigma(a_{mII}) = 0.133\%$, respectively.

The density of distribution of the first and second batch of the crystals with respect to the value a_m and their approximation by the Gauss function are presented in Fig. 4.19 and 4.20. The ORIGIN programme was used for statistical processing of the experimental data.

For reference samples of the supplied batches (for the first batch 3042 crystals, for the second batch 2476 crystals) the BTCP determined (in irradiation with the saturation dose) the values of the change of optical transmission δT_S, and the experimental density of distribution of these values and the approximation by the Gauss function are shown in Fig. 4.21 and 4.22.

The variation of optical transmission in saturation is characterised by the mean value and RMS deviation which equalled: for the first batch: $<\delta T_{S\,\text{I}}> = 8.8\%$ and $\sigma(\delta T_{S\,\text{I}}) = 3.2\%$; for the second batch: $<\delta T_{S\,\text{II}}> = 5.7\%$ and $\sigma(\delta T_{S\,\text{II}}) = 3.1\%$, respectively.

As indicated by Fig. 4.19–4.22, the experimental distribution of the photoelectric term and of the losses in optical transmission at saturation

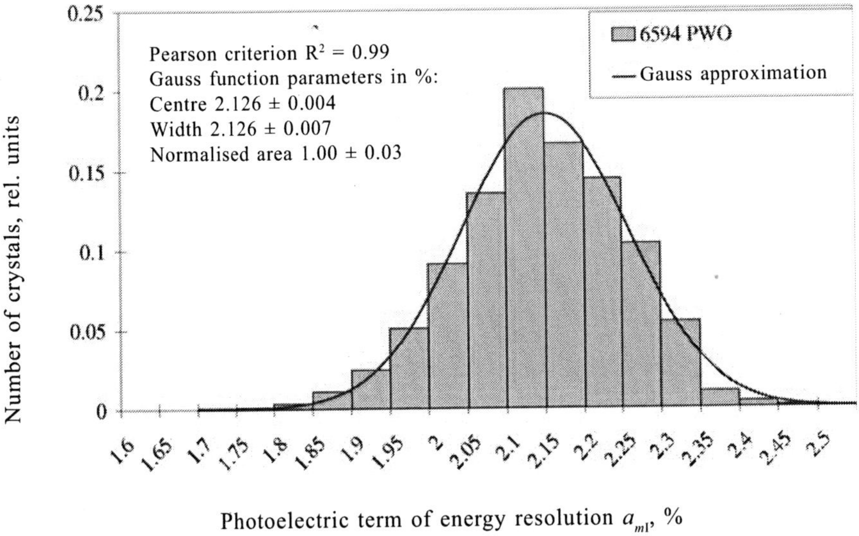

Fig. 4.19. The density of distribution of the crystals of the first batch on the basis of the value a_{ml} (%) and its approximation by the Gauss distribution.

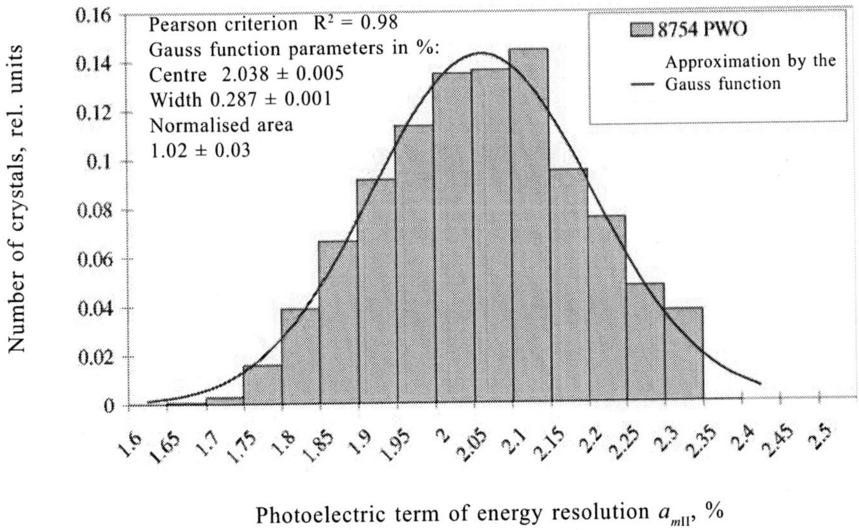

Fig. 4.20. The density of distribution of the crystals of the second batch on the basis of the value a_{mII} (%) and its approximation by the Gauss distribution.

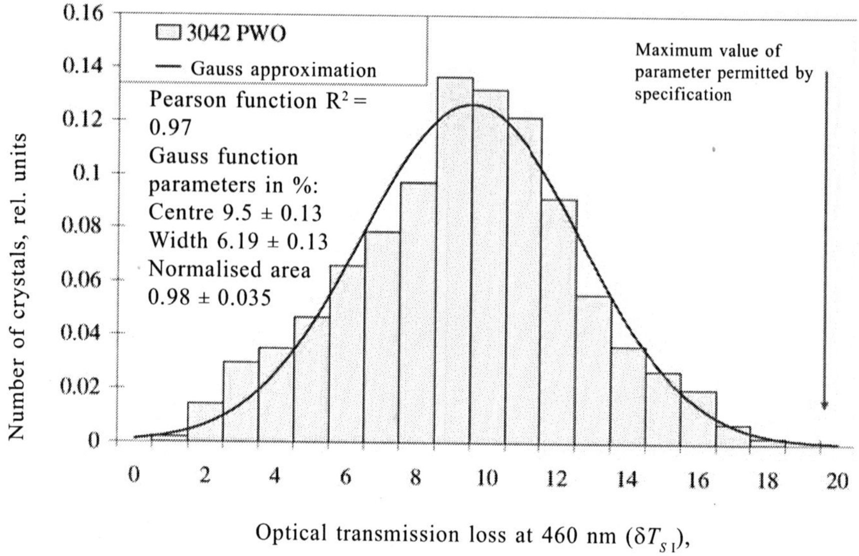

Fig. 4.21. The density of distribution of the first reference batch of crystals with respect to the value δT_S (%) and its approximation by the Gauss distribution.

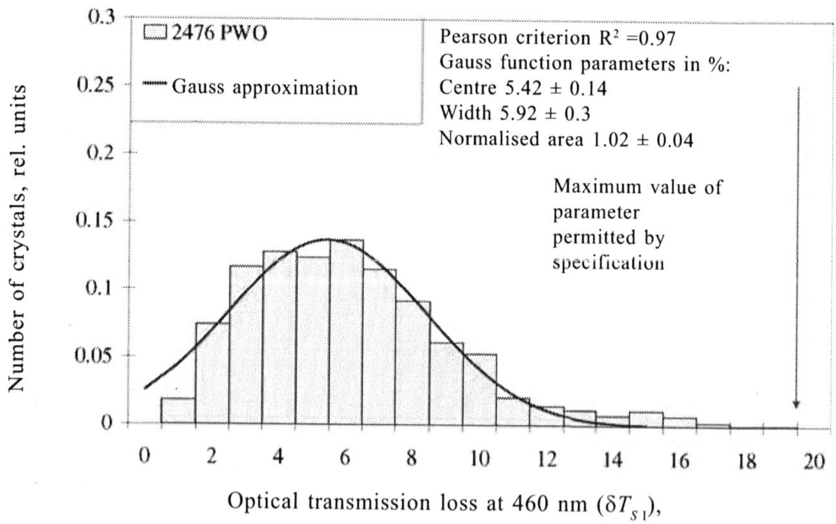

Fig. 4.22. The density of distribution of the second reference batch of crystals with respect to the value δT_S (%) and its approximation by the Gauss distribution.

are satisfactorily approximated by the normal law, taking into account that the values of the width of distribution, presented in the graphs, correspond to two RMS deviations. Therefore, in further stages when determining the parameters of the function of the investigated random quantities, it is permissible to use analytical methods instead of the relatively time-consuming numerical calculations.

For the reference sets of the crystals, the mean values of the photoelectric term were $<a_{c\,\mathrm{I}}> = 2.1\%$ and the RMS deviation $\sigma(a_{c\,\mathrm{I}}) = 0.11\%$ for the first batch, and $<a_{c\,\mathrm{II}}> = 2.0\%$, $\delta(a_{c\,\mathrm{II}}) = 0.14\%$ for the second batch of the crystals, respectively.

The parameters of linear regression and correlation dependence between the values of the photoelectric term in the reference batch a_c and the relative variation of the optical transmission T_S in irradiation with the saturation dose were determined using the measurement results.

The linear regression equations have the following parameters (Table 4.1 a,b) for the reference batches:

$$\delta T_{S\,\mathrm{I}} = 4.55\ a_{c\,\mathrm{I}} + 4.03 \qquad (4.3.1)$$

$$\delta T_{S\,\mathrm{II}} = 5.38\ a_{c\,\mathrm{II}} - 2.09 \qquad (4.3.2)$$

Because of the representative nature of the reference batches, the linear regression equation δT_S with respect to a_c, determined from the measured parameters of the statistical distributions and approximation by the normal law, are regarded as valid for both batches of the crystals.

The parameters of the correlation dependences a_c, δT_S were obtained for reference batches and applied to the complete batches:

- the sample correlation coefficients for the first batch $r_{\mathrm{I}} = 0.1$ and for the second batch $r_{\mathrm{II}} = 0.16$;
- the confidence ranges of the correlation coefficients: upper $\rho^h_{s\mathrm{I}} = 0.14$ and lower $\rho^l_{s\mathrm{I}} = 0.07$ for the first batch, and $\rho^h_{s\mathrm{II}} = 0.20$ and $\rho^l_{s\mathrm{II}} = 0.12$ for the second batch.

The results of measurements of δT_S at the saturation dose at the BTCP were compared with the results of measurements obtained with the working doses of the variation of optical transmission δT_V at the GIF CERN for 31 PWO crystals of the first batch and 20 PWO crystals in the second batch. All the crystals were produced by the same technology.

The mean value $\delta T_{W\,\mathrm{I}} = 3.4\%$ and the RMS deviation $\sigma(\delta T_{W\,\mathrm{I}}) = 0.82\%$ were determined for the first batch of the crystals and $\delta T_{W\,\mathrm{II}} = 2.2\%$ and $\sigma(\delta T_{W\,\mathrm{II}}) = 1.04\%$ for the second batch.

Relationships were determined between the distribution parameters of the crystals with respect to the variation of the optical transmission at the saturation dose and the working doses both batches. These dependences have the following form for the mean values:

$$\delta T_{W\,\mathrm{I}} = 0.17\ \delta T_{S\,\mathrm{I}} + 1.16 \qquad (4.4.1)$$
$$\delta T_{W\,\mathrm{II}} = 0.22\ \delta T_{S\,\mathrm{II}} + 0.25 \qquad (4.4.2)$$

The parameters of the linear regressions, obtained in examining six specimens from the first batch and 20 specimens from the second batch of the crystals at the saturation dose at the BTCP and the working dose at the GIF CERN, are in good agreement with the parameters of the dependences (4.1) and (4.2). As indicated by Fig. 4.23 and 4.24, the sample values of the correlation coefficients between the relative variations of transmission at the saturated and working doses are equal to 0.86 for the first batch and 0.89 for the second batch of the crystals. This is close to 1.

Therefore, it can be assumed that the parameters of the experimentally correlation (not determined in the experiments) between the distribution of the photoelectronic term in the batch of crystals a_m

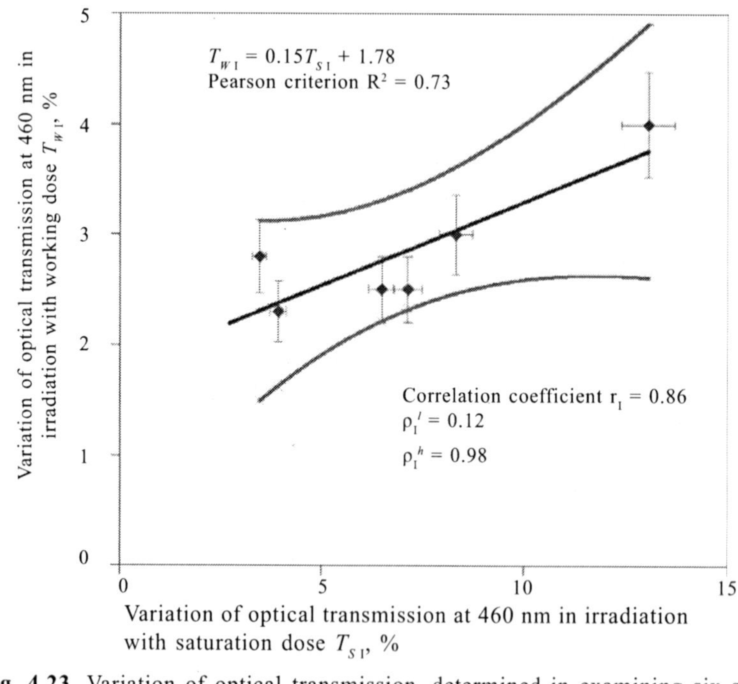

Fig. 4.23. Variation of optical transmission, determined in examining six specimens of the same crystals of the first batch at the saturation and working doses.

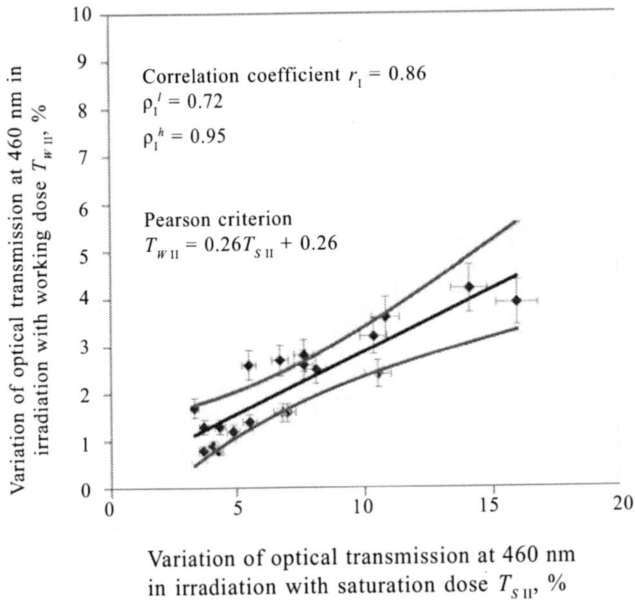

Fig. 4.24. Variation of optical transmission, determined in examining 20 specimens of the same crystals of the second batch at the saturation and working doses.

and the relative variation of transmission at the working dose δT_W can be represented by the correlation parameters between a_m and δT_S at the saturation dose for both batches i.e., it may be assumed that the confidence ranges ρ_S^h and ρ_S^l and the sample value of the correlation coefficient r_S can be used as the values of ρ_W^h, ρ_W^l and r_W, respectively.

The dependences between the variation of optical transmission δT_W and the relative variation of the photoelectronic term a_W for the working dose were determined for 15 crystals of the first batch and 10 crystals of the second batch. The experimental results, the derived regression equations and the parameters of the correlation relationships between δT_W and δa_w for both batches are presented in Fig. 4.25 and 4.26 and in Table 4.1 a, b. Physical investigations of radioluminescence in the PWO crystals show that the changes in the transmission and light yield are linked together by a proportional dependence. Therefore, to estimate the relative variation of the photoelectronic term δa_w, it was necessary to determine the value and error of the angular coefficient of the relationship between δa_W and $\delta T_W - k_W$, assuming that the correlation coefficients between δa_W, δT_W are equal to 1 for both batches.

On the basis of the determined characteristics and the assumptions, the parameters of distribution of the photoelectronic term for the working dose can be estimated on the basis of the following functional

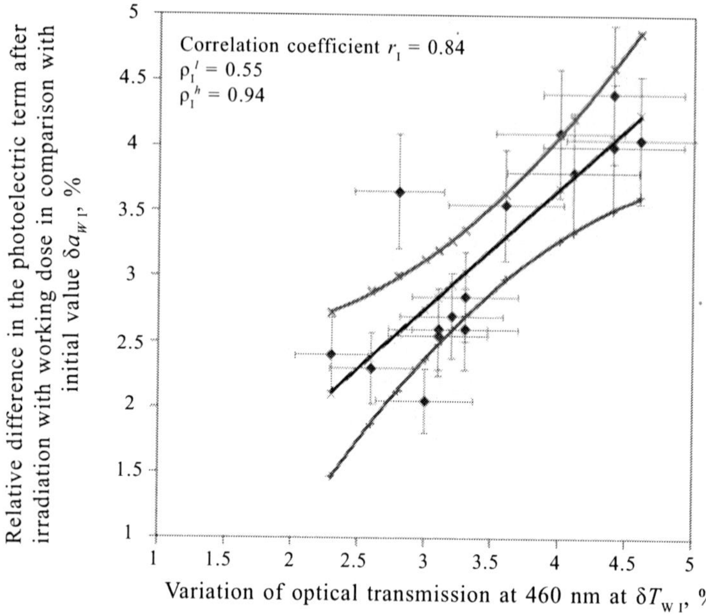

Fig. 4.25. The relationship between the variation of optical transmission $\delta T_{W I}$ and the relative variation of the photoelectric term $\delta a_{W I}$ in irradiation with the working dose of 15 PWO crystals for the first batch.

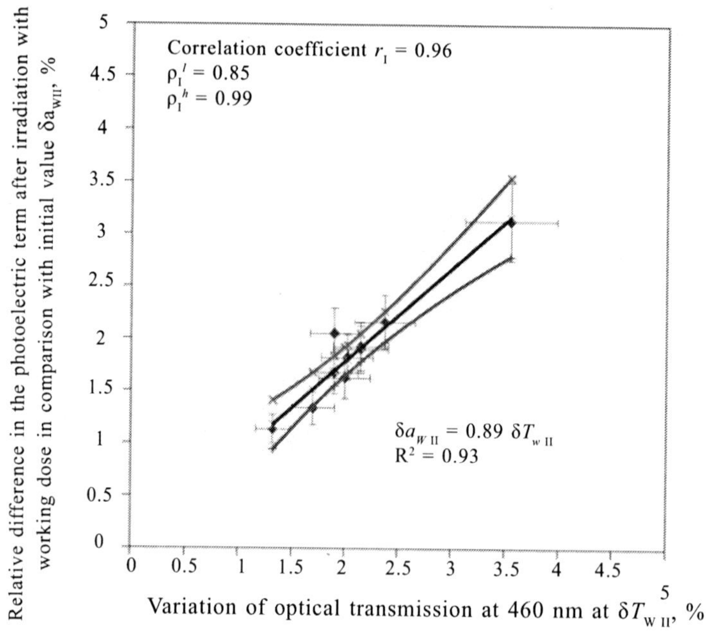

Fig. 4.26. The relationship between the variation of optical transmission $T_{W II}$ and the relative variation of the photoelectric term $a_{W II}$ in irradiation with the working dose of 10 PWO crystals for the second batch.

Table 4.1a. Investigated characteristics of the first batch of PWO crystals for obtaining statistical estimates of the parameters of radiation hardness

No.	Data set	Number	Measured parameters	Analysed parameters	Correlation coefficient r, upper ρ^h and lower ρ^l confidence ranges	Linear regression equation
1	Reference set	3042	N_{phe} $\delta T_{S\,I}$	Correlation cor ($\delta a_{m\,I}$; $\delta T_{S\,I}$) Linear regression $\delta a_{m\,I}$; $\delta T_{S\,I}$	$r_{S\,I} = 0.1$ $\rho^l_{S\,I} = 0.07$ $\rho^h_{S\,I} = 0.14$	$\delta T_{S\,I} = 4.55 \cdot a_{m\,I} + 4.03$
2	Experimental crystals	31	$\delta T_{W\,I}$	$\delta T_{W\,I}$; $W(\delta T_{W\,I})$	$\delta T_{W\,I} = 3.4$ $W(\delta T_{W\,I}) = 1.64$	
		15	$\delta T_{W\,I}$ $a_{W\,I}$	Correlation cor ($\delta a_{W\,I}$; $\delta T_{W\,I}$) Linear regression $\delta a_{W\,I}$; $\delta T_{W\,I}$	$r_{W\,I} = 0.84$ $\rho^l_{W\,I} = 0.55$ $\rho^h_{W\,I} = 0.94$	$\delta a_{W\,I} = 0.92 \cdot T_{W\,I}$ Pearson coefficient $R^2 = 0.70$
3	Experimental crystals	6	$\delta T_{S\,I}$ $\delta T_{W\,I}$	Correlation cor ($\delta T_{W\,I}$; $\delta T_{S\,I}$) Angular correlation coefficient	$r_I = 0.86$ $\rho^l_{S\,I} = 0.12$ $\rho^h_{S\,I} = 0.98$	$\delta T_{W\,I} = 0.15 + \delta T_{S\,I} + 1.8$ Pearson coefficient $R^2 = 0.73$

Table 4.1b. Investigated characteristics of the second batch of PWO crystals for obtaining statistical estimates of the parameters of radiation hardness

No.	Data set	Number	Measured parameters	Analysed parameters	Correlation coefficient r, upper ρ^h and lower ρ^l confidence ranges	Linear regression equation
1	Reference set	2476	N_{phe} $\delta T_{S\,II}$	Correlation cor ($\delta a_{m\,II}$; $\delta T_{S\,II}$) Linear regression $\delta a_{m\,II}$; $\delta T_{S\,II}$	$r_{S\,II} = 0.16$ $\rho^l_{S\,II} = 0.12$ $\rho^h_{S\,II} = 0.20$	$\delta T_S = 5.38 \cdot a_{m\,II} - 2.09$
2	Experimental crystals	20	$\delta T_{W\,II}$	$\delta T_{W\,II}$; $W(\delta T_{W\,II})$	$\delta T_{W\,II} = 2.2$ $W(\delta T_{W\,II}) = 2.08$	
		10	$\delta T_{W\,II}$ $a_{W\,II}$	Correlation cor (δa_W; δT_W) Linear regression $\delta a_{W\,II}$; $\delta T_{W\,II}$	$r_{W\,II} = 0.96$ $\rho^l_{W\,II} = 0.85$ $\rho^h_{W\,II} = 0.99$	$\delta a_{W\,II} = 0.89 \cdot T_{W\,II}$ Pearson coefficient $R^2 = 0.93$
3	Experimental crystals	20	$\delta T_{S\,II}$ $\delta T_{W\,II}$	Correlation cor ($\delta T_{W\,II}$; $\delta T_{S\,II}$) Angular correlation coefficient	$r_{II} = 0.89$ $\rho^l_{S\,II} = 0.72$ $\rho^h_{S\,II} = 0.95$	$\delta T_{W\,II} = 0.26 + \delta T_{S\,II} + 0.26$ Pearson coefficient $R^2 = 0.73$

dependence with the distribution of the photoelectronic term in two batches of the crystals:

$$a_W = (1-k_W\, \delta T_W)a_M \tag{4.5}$$

where k_W is a dimensionless quantity, and the remaining quantities are in %, on the condition that

$$r(\delta T_S, a_m) = r(\delta T_W, a_m) \tag{4.6}$$

For the two batches of the PWO crystals for the radiation conditions in the electromagnetic calorimeter with a working dose, equations (4.5) and (4.6) were used to calculate the parameters characterising the radiation hardness of the crystals. These are mean values of a_W and the width W (a_W) of the distributions of the crystals with respect to the value of the photoelectronic term in irradiation with the working dose with the estimate of the error of the resultant values of (a_W) and $\sigma(W)$, respectively.

Since the reliability of the calculated estimates of a_W and $W(a_W)$ is associated with the extent of the experimental investigations of the variation of the optical transmission of the photoelectronic term of the working dose, the lower and upper boundaries of the confidence ranges $a_W^{\,l}$, $a_W^{\,h}$ and $W^l(a_W)$, $W^h(a_W)$ respectively were determined for the mean values and width with the probability of 0.95.

The estimates of the values of these characteristics of the statistical distribution of the photoelectronic term for the first and second batches of the crystals at the working dose are presented in Tables 4.2*a* and *b*. The values of a_m and $W(a_m)$ and the error of determination of these values $\sigma(a_m)$ and $\sigma(W)$ were obtained from the results of certification measurements of the values of the photoelectronic yield of the crystals of the two batches at the BTCP.

Tables 4.2 *a*, *b* also show the estimates of the number of crystals (in percent) in technological batches with the value of the electronic term exceeding the maximum value of 2.3% permitted by the supply specifications for the statistical parameters of the initial distribution a_m and the resultant estimates of the distribution parameters a_w, $a_W^{\,l}$ and $a_W^{\,h}$.

The calculation showed that the initial distribution of the 6594 PWO crystals of the first batch corresponds to the mean value of the photoelectronic term of 2.126±0.004 and the distribution width of 0.216±0.007. For the 8754 crystals of the second batch, the mean value and width are equal to 2.038±0.005 and 0.287±0.010, respectively. Under the effect of radiation at the dose corresponding to the mean dose loading in the ECAL CMS, with the probability of $P = 0.95$, the centre of distribution is displaced to the value 2.192±0.004, and the width of distribution to 0.220±0.008 for the first batch of the crystals and for the second batch of the crystals to 2.077±0.004 and

Table 4.2a. Estimates of the characteristics of the statistical distribution of the photoelectronic term at the working dose for the first batch of the crystals

Number of crystals	Prior to irradiation	After irradiation with working dose
Parameters	$a_{m\,\mathrm{I}} = 2.126$ $\mathrm{W}(a_{m\,\mathrm{I}}) = 0.216$	$a_{W\,\mathrm{I}} = 2.192$ $\mathrm{W}(a_{W\,\mathrm{I}}) = 0.22$
Error of determination of parameters	$\sigma(a_{m\,\mathrm{I}}) = 0.0035$ $\sigma\,(W(a_{m\,\mathrm{I}}) = 0.007$	$\sigma(a_{W\,\mathrm{I}}) = 0.0039$ $\sigma(\mathrm{W}(a_{W\,\mathrm{I}})) = 0.008$
Confidence ranges of parameters		$a_{W\,\mathrm{I}}{}^{l}=2.188$, $a_{W\,\mathrm{I}}{}^{h}= 2.196$ $\mathrm{W}^{l}(a_{W\,\mathrm{I}}) = 0.213$ $\mathrm{W}^{h}(a_{W\,\mathrm{I}}) = 0.227$
Number of crystals (in%) with the value of the term higher than the maximum value permitted by specification (2.3%)	$Q(a_{m\,\mathrm{I}}) = 3.09\%$	$Q(a_{W\,\mathrm{I}}) = 11.14\%$ $Q(a^{l}{}_{W\,\mathrm{I}}) = 9.72\%$ $Q(a^{h}{}_{W\,\mathrm{I}}) = 12.6\%$

Table 4.2b. Estimates of the characteristics of the statistical distribution of the photoelectronic term at the working dose for the second batch of the crystals

Number of crystals	Prior to irradiation	After irradiation with working dose
Parameters	$a_{m\,\mathrm{II}} = 2.038$ $W(a_{m\,\mathrm{II}}) = 0.287$	$a_{W\,\mathrm{II}} = 2.077$ $W(a_{W\,\mathrm{II}}) = 0.311$
Error of determination of parameters	$\sigma(a_{m\,\mathrm{II}}) = 0.005$ $\sigma(\mathrm{W}(a_{m\,\mathrm{II}}) = 0.01$	$\sigma(a_{W\,\mathrm{II}}) = 0.004$ $\sigma(W(a_{W\,\mathrm{II}})) = 0.021$
Confidence ranges of parameters		$a_{W\,\mathrm{II}}{}^{l}=2.073$, $a^{h}{}_{W\,\mathrm{II}}= 2.081$ $\mathrm{W}^{l}(a_{W\,\mathrm{II}}) = 0.290$ $\mathrm{W}^{h}(a_{W\,\mathrm{II}}) = 0.332$
Number of crystals (in%) with the value of the term higher than the maximum value permitted by specification (2.3%)	$Q(a_{m\,\mathrm{II}}) = 0.25\%$	$Q(a_{W\,\mathrm{II}}) = 3.22\%$ $Q(a^{l}{}_{W\,\mathrm{II}}) = 1.62\%$ $Q(a^{h}{}_{W\,\mathrm{II}}) = 4.97\%$

0.311±0.021, respectively. The probability functions of the distribution of the first and second batch of the PWO crystals at the exit (–) and after irradiation with a working dose (d = 6 Gy) for the lower (---) and upper (-.-.-) boundaries of the confidence range with the probability P = 0.95 are presented in Fig. 4.27 and 4.28.

Thus, for the PWO crystals, design for placing in the barrel of the electromagnetic calorimeter, experiments were carried out to determine the variation of the parameters of the statistical distribution of the photoelectric term of energy resolution in the irradiation conditions simulating the working values of the power of the dose in the central part of the calorimeter. The proposed method is used for taking

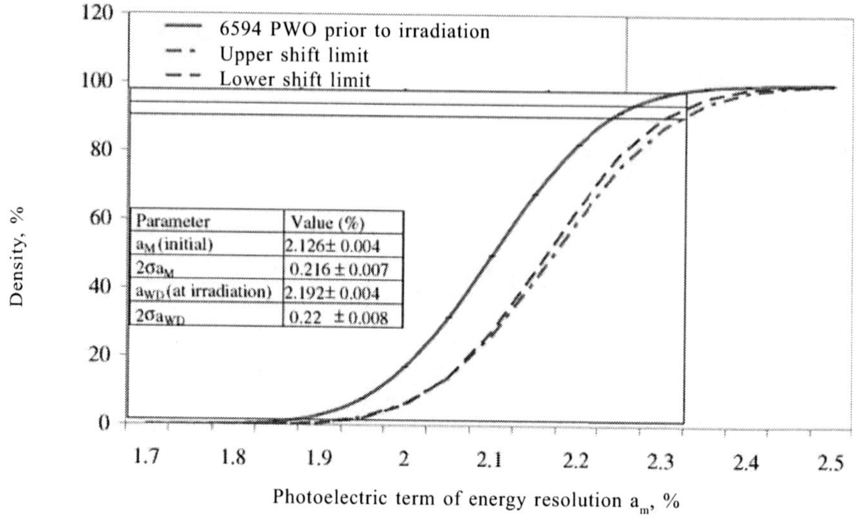

Fig. 4.27. The function of the probability of distribution of the first batch of PWO crystals at the exit (–) and after irradiation with a working dose (d = 6 Gy) for the lower (----) and upper (-.-.-) shift limits with the probability of P = 0.95.

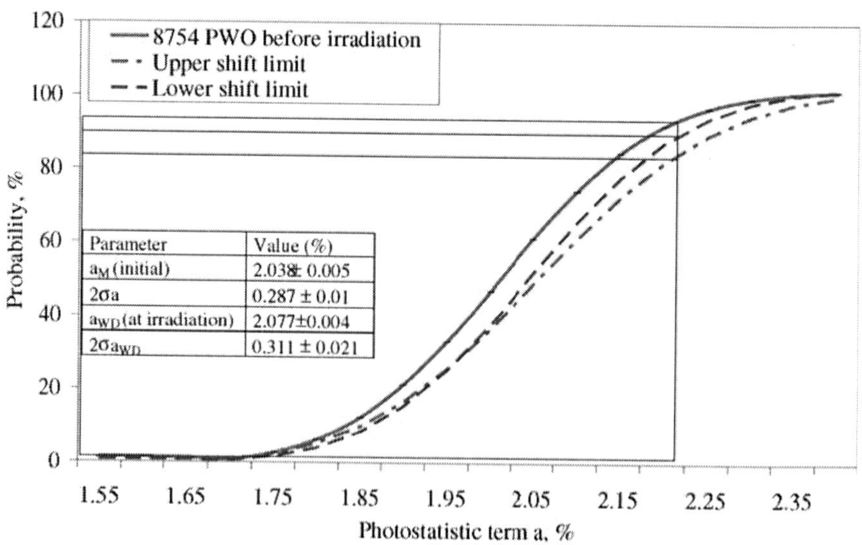

Fig. 4.28. The function of the probability of distribution of the second batch of PWO crystals at the exit (–) and after irradiation with a working dose (d = 6 Gy) for the lower (----) and upper (-.-.-) shift limits with the probability of P = 0.95.

decisions regarding the application of every produced scintillation PWO element in a specific region of the electromagnetic calorimeter in the CMS project, depending on the dose.

Conclusions

We hope that you enjoyed reading the book and found most of it or parts of it interesting. In any case, we are grateful to you for this. We also hope that you are found something interesting and new in this book.

The book should be of interest to a wide range of scientists, engineers and technologists, concerned with the problems of materials science and the application of scintillators. Experience with the work on a large international projects, described in the book, will be used to utilize the results in growing other crystals and also to transfer to the reader the satisfaction from the work and the results obtained by the group of scientists and experts from different countries. The crystals are a very interesting object for investigations but the process of development of technology for industrial mass production of the crystals is also interesting. This process can be described as innovative. It includes the latest achievements and new concepts from related disciplines and in many cases from greatly differing areas of science and technology. In particular, these are the features which the authors have tried to present to the readers.

Progress does not stagnate. Experts in the area of synthetic crystals will face new tasks and problems: for example, the search for and production of crystals with unique properties for recording weak interactions in experiments concerned with the search for dark matter and for detector systems of medical tomograms of a new generation. Undoubtedly, because of the scale and importance of these tasks, it will be necessary to find new approaches to the development of technological processes. Therefore, the experience with the organisation of the mass production of lead tungstate scintillator crystals, described in the book, may prove useful to experts in the appropriate disciplines. Artificial crystals are equally art and science, high technology and business and, finally, in our experience, the possibility of communicating with interesting people. One of our friends noted that the crystals, like people, are produced in the same manner but they greatly differ. Therefore, there are no limits to developing evrn more perfect crystalline materials.

The authors are grateful for scientific coopreation to teams from BTCP, CERN, MNNTs (International Research and Technology Center), NIIYaP (Nuclear Physics Research Institute), IFVE (High Energy

Physics Institute),, MISiS (Moscow Institute of Steel and Alloys) and personally to M.V. Korzhik, V.A. Kachanov, A.E. Losovitskii, V.D. Ligun, P. Lecoq, M. Lebeau, E. Auffray-Hillemans, G.Yu. Drobyshev, R.F. Zuevskii, O.V. Misevich, A.V. Sin'govskii and A.A. Fedorov, and would also like to mention the significant contribution to this project of scientits and experts who, sadly, are no longer with us: V.L. Kostylev, A.A. Blistanov and O.N. Kovalev.

Yurii Sergeevich Kuz'minov

We are also saddened to report that one of the authors of the book, Prof. Kuz'minov, died in July 2009. We shall all miss him very much

The Publishers
Cambridge International Science Publishing

ter A., Alves H., Bohm M., et al., Inorganic scintillators and their application, Ed. V. Mikhailin, Moscow State University, 2000, p. 128–136.

41. The role of the defect states in the creation of the intrinsic WO_4^{3-} centers in $PbWO_4$ by sub-band excitation., Hofstaetter A., Korzhik M. V., Laguta V. V., et al., Radiation Measurements, 2001. Vol. 33, p. 533–536.

42. Influence of Mo impurity on the spectroscopic and scintillation properties of $PbWO_4$ crystals., Böhm M., Borisevich A.E., Drobychev G.Yu. et al, Phys. stat. sol. (a), 1998, V. 167, No.1, p. 243–252.

43. Photoinduced oxygen vacancy related centers in $PbWO_4$: electron spin resonance and thermally stimulated luminescence study, V. Laguta, M. Martini, A. Vedda, et al., Rad. Eff and Def. in Solids. 2002, Vol. 157, p. 1025–1031.

44. Electron traps related to oxygen-vacancies in $PbWO_4$, V. Laguta, M. Martini, A. Vedda, et al., Accepted for publication in Phys. Rev., 2003.

45. Observation of dipole complexes in $PbWO_3$:La^{3+} singler crystals, Han B., Feng X., Hu G., et al., J. Appl. Phys, 1998. Vol. 84, p. 2831–2834.

46. Huang H., Feng X., Man Z., Impedance spectroscopy analysis of La-doped $PbWO_4$ single crystals, J. Applied Physics. 2003, Vol. 93. p. 421–425.

47. Коржик М.В. Физика сцинтилляторов на основе кислородных монокристаллов. – Минск, БГУ, 2003,102 с.

48. Improvement in radiation hardness of $PbWO_4$ scintillationg crystals by La-doping, M. Kobayashi, Y. Usuki, M. Ishii, et al., Nucl. Instr. Meth. Phys. Res. A, 1998, Vol. 404, p. 149–156.

49. Decay kinetics and thermoluminescence of $PbWO_4$, M. Nikl, K. Nitsch, K. Polak, et al., Phys. Stat. Sol. (b) - 1996, Vol. 195, p. 311–323.

50. Decay kinetics and thermoluminescence of $PbWO_4$:La^{3+}, Nikl M., Bonacek P., Nitsch K., et al., Applied Phys. Lett, 1997, Vol. 71, p. 3755–3757.

51. A study on the properties of lead tungstate crystals, Zhu R. Y., Ma D. A., Newman H. B., et al., Nucl. Instr. Meth. Phys. Res. A, 1996, Vol. 376, p.319–334.

52. Thermally stimulated luminescence of $PbWO_4$ crystals, Martini M., Rosetta E., Spinolo G., et al., J. Lumin, 1997, Vol. 72–74, p. 689–690.

53. Radiation induced formation of color centers in $PbWO_4$ single crystals., Nikl M., Nitsch K., Baccaro S., et al., J. Appl. Phys, 1997, Vol. 82, p.5758–5762.

54. Weightman P., Henderson B., Dugdale D.E., An EPR Study of Divacancy Centres in $CaWO_4$, Phys. Stat. Sol. (b), 1973, Vol. 58, p.321–330.

55. Lead tungstate scinatillation material for precision electromagnetic calorimetry in new generation accelerators, Korzhik M.V., Annenkov A.N., Drobyshev G.Yu., et al., Selected studied of the Belorussian National University, vol. 4, Physics, Minsk, 2001, p. 323–342.

56. Picosecond and nanosecond time-resolved study of luminescence and absorption of $CdWO_4$ and $PbWO_4$, Williams R. T., Yochum H. M., Ucer K. B, et al., Inorganic scintillators and their applications, Ed. V. Mikhailin, Moscow State University, 2000, p. 336–341.

57. Luminescence and transient absorption of doped $PbWO_4$ scintillator crystals, Millers D., Chernov S., Grigorieva L., et al., Inorganic scintillators and their applications, Ed. V. Mikhailin, Moscow State University, 2000, p. 613–618.

58. Color center production in $PbWO_4$ crystals by UV light exposure, I. Dafinei, B. Borgia, F. Cavallari, et al., Inorganic Scintillators and Their Applications, Ed. Yin Zhiwen, Li Peijun, Feng Xiqi, Xue Zhilin, Shanghai, September 1997, p. 219–222.

59. Doping $PbWO_4$ with different ions to increase the light yield, Kobayashi M., Usuki Y., Ishi M., Nikl M., Nucl. Intr. Meth. Phys. Res. A, 2002, Vol. 486, p.

170–175.

60. Influence of PbF2 doping on scintillation properties of $PbWO_4$ single crystals, Liu X., Hu G., Feng X., et al., Phys. Stat. Sol. (a), 2002, Vol. 190, p. 1–3.
61. Compact Muon Solenoid Experiment, CMS Collaboration, Geneva, Switzerland, 1998, http://cmsinfo.cern.ch, Brochures, IntroToCMS.pdf.
62. Lead tungstate ($PbWO_4$) scintillators for LHC EM-calorimetry, Lecoq P., Dafinei I., Auffray E., Annenkov A., et al., Nucl. Instr. Meth. Phys. Res. A, Vol. 265. p. 291–298.
63. Lecoq P., Korzhik M., Scintillator developments for high energy physics and medical imaging, IEEE Trans. Nucl. Sci, 2000, Vol. 47, p. 1311–1314.
64. Significant increase in fast scintillation component from $PbWO_4$ for PET by optimum annealing, Kobayashi M., Usuki Y., Ishii M., Itoh M., Inorganic scintillators and their applications. 7th International Conference on Inorganic scintillators and industrial applications. Book of Abstracts, Valencia, Spain, 8–12 September 2003, p. 47.
65. Improved light yield of lead tungstate scintillators, Annenkov A., Borisevitch A., Hofstaetter A., et al., Nucl. Instr. Meth. in Physics Research A, 2000, Vol. 450, p. 71–74.
66. Efficient medium speed $PbWO_4$:Mo,Y scintillator, Nikl M., Bohacek P., Vedda A., et al., Physics. Stat. Solid.(a), 2001, Vol. 186, p. 1–3.
67. Stability of scintillation light yield under small radiation doses, Fedorov A.A., Pavlenko V.B., Korzhik M. V. et al.,, Radiation Measurements, 1996, Vol. 26, p. 215.
68. Radiation hardness of doped $PbWO_4$, Kobayashi M., Usuki Y., Ishii M. et al., Inorganic scintillators and their applications, Ed. V. Mikhailin, MSU, 2000, p. 137–146.
69. Suppression of the radiation damage in lead tungstate scintillation crystal., Annenkov A., Auffray E., Borisevich A., et al., Nucl. Instr. Meth. Phys. Res. A, 1999, Vol. 426, p.486–490.
70. On the Mechanism of Radiation Damage of Optical Transmission in Lead Tungstate Crystal., Annenkov A.N., Borisevich A.E., Drobychev G.Yu., et al., Phys. Stat. Sol. (a), 2002, Vol. 191, Issue 1, p. 277–290.
71. Technical conditions 2624-003-17611446, Charge for lead tungstate single crystals, OSCh 10-3, Standards Publishing House, Moscow, 1999.
72. Production of specified raw materials for mass manufacturing of radiation hard scintillation materials., Dossovitski A.E., Mikhlin A.L., Annenkov A.N., Nucl. Instr. Meth. Phys. Res. A, 2002, Vol. 486, p. 98–101.
73. Influence of stoichiometry on the optical properties of lead tungstate crystals, Belsky A.N., V.V.Mikhailin, Annenkov A.N., et al., Chem. Phys. Lett., 1997, Vol. 277, p. 65–70.
74. Russian Federation Parent 2132417, MCI 6 S30V 15, 04, 29, 32. A method of proudcing lead tungstate scintillation single crystals, Anneakov A.N., Korzhik M.V., Kostylev V.L., Ligun V.D., Claim 22.01.1998; Publ. 27.06.1999, State Inventions Register of Russian Federation, Moscow, 1999.
75. Russian Federation Parent 2164562, MCI 6 S30V 15, 04, 29, 32. A method of proudcing lead tungstate scintillation single crystals, Annenkov A.N., Korzhik M.V., Kostylev V.L., Ligun V.D., Claim 18.02.2000; Publ. 27.04.2001, State Inventions Register of Russian Federation, Moscow, 2001.
76. Dmitriev V.A., High-temperature failure of platinum metals and alloys, Ruda i Metally Publishing House, Moscow, 2003.
77. Methodology of Certification of Scintillators for Large Scale Detectors, Auffray

E., Annenkov A.N., Drobychev G.Yu., et al., SCINT'99. The Fifth International Conference on Inorganic Scintillators and Their Applications, August 16–20, 1999, Moscow, Russia, p. 212–217.

78. Develorment of Uniformisation Procedure for the $PbWO_4$ Crystals of the CMS Electromagnetic Calorimeter, Auffray E., Davies G.J., Lebeau M. et al., Preprint – CMS Note 2001, 004, CERN, Geneva, Switzerland, 2001.

79. Certifying Procedures for Lead Tungstate Crystal Parameters During Mass Production for the CMS ECAL., Auffray E., Drobychev G.Yu., Korzhik M.V., et al., IEEE'98 Abstr. Toronto, Canada, November 8–14, 1998, No. 20–31, p. 6.

80. Studies and Proposals for an Automatic Crystal Control System, Drobychev G., Korzhik M., Peigneux J.P., et al., Preprint – CMS TN No.036 – CERN, Geneva, Switzerland, 1997, 38 p.

81. Analysis of ACCOS System results reproducibility and results of first pre-mass production PWO certification, Drobychev G.Yu., Auffray E., Korzhik M.V., et al., Preprint LAPP-EXP-99.07. LAPP, Annecy-le-Vieux, 1999.

82. Performance of ACCOS, an Automatic Crystal Quality Control System for the PWO crystals of the CMS calorimeter, Auffray E., Chevenier G., Drobychev G., et al., Nucl. Instr. and Meth. in Phys. Res. A, 2001, Vol. 456, Issue 3, p. 325–341.

83. Equipment and Methods for Rapid Analysis of PWO Full Size Scintillation Crystals Radiation Hardness at Mass Production, Annenkov A., Auffray E., Drobychev G. et al., IEEE Trans. on Nucl. Sci, 2001, Vol. 48, No. 4 – p. 1177–1181.

84. Results of PWO radiation hardness optimization, Drobychev G.Yu., Annenkov A.N., Auffray-Hillemans E., et al.,Preprint CMS Note 1999-062, CERN, Geneva, Switzerland, 1999.

85. Radiation Hardness of Mass Produced PWO Crystals., Drobychev G., Annenkov A., Auffray-Hillemans E. et al., IEEE'2000. Nucl. Sci. Symp. Conf. Rec., Lyon, France, October 15–20, 2000, NSS–57, 4 p.

86. $PbWO_4$ crystals radiation hardness test setup at the CERN General Irradiation Facility, Peigneux J-P., Singovski A., Fedorov A, et al.,Preprint - CMS Note 1999, 061. CERN, Geneva, Switzerland, 1999.

87. Influence of the distribution of PWO crystal radiation hardness on electromagnetic calorimeter performance, Drobychev G.Yu., Borisevich A.E., Korjik M.V. et al., Nuclear Inst. and Meth. in Phys. Res., A, 2002, Vol. 486, Issue 1–2, p. 116–120.

Index